Lecture Notes in Computer Science 3534

Commenced Publication in 1973
Founding and Former Series Editors:
Gerhard Goos, Juris Hartmanis, and Jan van Leeuwen

T0223554

Stefano Spaccapietra Esteban Zimányi (Eds.)

Journal on
Data
Semantics III

 Springer

Editor-in-Chief

Stefano Spaccapietra
Database Laboratory, EPFL
1015 Lausanne, Switzerland
E-mail: stefano.spaccapietra@epfl.ch

Volume Editor

Esteban Zimányi
Université Libre de Bruxelles
Laboratory of Computer and Network Engineering
CP 165/15, 50 av. F.D. Roosevelt, 1050 Brussels, Belgium
E-mail: ezimanyi@ulb.ac.be

Library of Congress Control Number: 2004117338

CR Subject Classification (1998): H.2, H.3, I.2, H.4, C.2

ISSN   0302-9743
ISBN-10   3-540-26225-3 Springer Berlin Heidelberg New York
ISBN-13   978-3-540-26225-1 Springer Berlin Heidelberg New York

Springer is a part of Springer Science+Business Media

springeronline.com

© Springer-Verlag Berlin Heidelberg 2005
Printed in Germany

Typesetting: Camera-ready by author, data conversion by Scientific Publishing Services, Chennai, India
Printed on acid-free paper   SPIN: 11496168   06/3142   5 4 3 2 1 0

The LNCS Journal on Data Semantics

Computerized information handling has changed its focus from centralized data management systems to decentralized data exchange facilities. Modern distribution channels, such as high-speed Internet networks and wireless communication infrastructures, provide reliable technical support for data distribution and data access, materializing the new, popular idea that data may be available to anybody, anywhere, anytime. However, providing huge amounts of data on request often turns into a counterproductive service, making the data useless because of poor relevance or inappropriate level of detail. Semantic knowledge is the essential missing piece that allows the delivery of information that matches user requirements. Semantic agreement, in particular, is essential to meaningful data exchange.

Semantic issues have long been open issues in data and knowledge management. However, the boom in semantically poor technologies, such as the Web and XML, has boosted renewed interest in semantics. Conferences on the Semantic Web, for instance, attract crowds of participants, while ontologies on their own have become a hot and popular topic in the database and artificial intelligence communities.

Springer's LNCS Journal on Data Semantics aims at providing a highly visible dissemination channel for most remarkable work that in one way or another addresses research and development on issues related to the semantics of data. The target domain ranges from theories supporting the formal definition of semantic content to innovative domain-specific application of semantic knowledge. This publication channel should be of highest interest to researchers and advanced practitioners working on the Semantic Web, interoperability, mobile information services, data warehousing, knowledge representation and reasoning, conceptual database modeling, ontologies, and artificial intelligence.

Topics of relevance to this journal include:

- semantic interoperability, semantic mediators
- ontologies
- ontology, schema and data integration, reconciliation and alignment
- multiple representations, alternative representations
- knowledge representation and reasoning
- conceptualization and representation
- multimodel and multiparadigm approaches
- mappings, transformations, reverse engineering
- metadata
- conceptual data modeling
- integrity description and handling
- evolution and change
- web semantics and semistructured data

- semantic caching
- data warehousing and semantic data mining
- spatial, temporal, multimedia and multimodal semantics
- semantics in data visualization
- semantic services for mobile users
- supporting tools
- applications of semantic-driven approaches

These topics are to be understood as specifically related to semantic issues. Contributions submitted to the journal and dealing with semantics of data will be considered even if they are not within the topics in the list.

While the physical appearance of the journal issues looks like the books from the well-known Springer LNCS series, the mode of operation is that of a journal. Contributions can be freely submitted by authors and are reviewed by the Editorial Board. Contributions may also be invited, and nevertheless carefully reviewed, as in the case for issues that contain extended versions of best papers from major conferences addressing data semantics issues. Special issues, focusing on a specific topic, are coordinated by guest editors once the proposal for a special issue is accepted by the Editorial Board. Finally, it is also possible that a journal issue be devoted to a single text.

The journal published its first volume in 2003 and its second volume at the beginning of 2005. This is the third volume; the first one to be a special issue devoted to a specific theme. We are very grateful to Prof. Esteban Zimányi, from the Université Libre de Bruxelles, for accepting the load of organizing this special issue. Two other volumes are due to appear in 2005, and will be followed in 2006 by a special issue on Emergent Semantics.

The Editorial Board comprises one Editor-in-Chief (with overall responsibility) and several members. The Editor-in-Chief has a four-years mandate to run the journal. Members of the board have a three-years mandate. Mandates are renewable. More members may be added to the board as appropriate.

We are happy to welcome you to our readership and authorship, and hope we will share this privileged contact for a long time.

Stefano Spaccapietra
Editor-in-Chief
http://lbdwww.epfl.ch/e/Springer/

JoDS Volume 3 — Special Issue on Semantic-Based Geographical Information Systems

Geographical information has been established as a fundamental and strategic component of today's decision-support systems. Geographical information systems (GISs) have been successfully used in many diverse application domains, from land management to atmospheric and spatial observation, from history preservation and archaeology to biodiversity. However, new applications ask for enriching the semantics associated with geographical information in order to support a wide variety of tasks including data integration, interoperability, knowledge reuse, knowledge acquisition, knowledge management, spatial reasoning and many others. Examples of such semantic issues are temporal and spatiotemporal data management, 3D manipulation, spatial granularity, multiple resolutions, multiple representations, fuzzy and ambiguous geographic information, the relationship between geographic and physical concepts, and identity of geographic objects through time.

In addition, recent years have witnessed many technological developments that have radically changed how we understand information processing. Data warehouses and OLAP systems have evolved as a fundamental approach for developing advanced decision-support systems. This led to improved data mining techniques allowing us to extract semantics from raw data. Furthermore, the success of the Internet generated a paradigm shift in distributed information processing leading to the area of the Semantic Web, in which semantics is the fundamental component for achieving communication both for humans and applications. At the same time, mobile and wireless computing have entered everyone's life through dedicated devices leading to location-based services. Finally, Grid computing, a paradigm enabling applications to integrate computational and information resources managed by diverse organizations in widespread locations, pushes the frontier of global interoperability. The fact that all these recent developments are entering the geographic domain increases the importance of the elicitation of the semantics of geographical information.

The papers in this special issue address many of the topics mentioned above. They all provide different insights about the multiple benefits that can be obtained by envisioning GISs from a new semantic perspective. As this is a relatively new domain, these papers open many new research directions that need to be addressed in future work. This research will definitely have a huge impact on the next generation of GIS applications and tools.

March 2005

Esteban Zimányi
Special Issue Editor

Reviewers for the Special Issue

I would like to thank all the reviewers for their excellent work evaluating the papers. Without their commitment the publication of this special issue of the JoDS journal would never have been possible.

Andreas Abecker
Thomas Barkowsky
Claudia Bauzer Medeiros
Yvan Bédard
Dave Bennett
Michela Bertolotto
Alex Borgida
Alain Bouju
Bénédicte Bucher
Barbara Catania
Jean-Paul Cheylan
Eliseo Clementini
Christophe Claramunt
Isabel F. Cruz
Paolino Di Felice
Jean-Paul Donnay
David W. Embley
Antony Galton
Nicola Guarino
Diansheng Guo
Stephen Hirtle
Jerry R. Hobbs
Gary Hunter
Panos Kalnis
Zdenek Kouba
Werner Kuhn
Alberto H.F. Laender
Maurizio Lenzerini
Sergio Luján-Mora

David Mark
Daniel R. Montello
Bernard Moulin
Pedro R. Muro-Medrano
Leo Obrst
Christine Parent
Maria Teresa Pazienza
Donna Peuquet
Elaheh Pourabbas
Sham Prasher
Sudha Ram
Stéphane Roche
Marco Schorlemmer
Shashi Shekhar
Barry Smith
Stefano Spaccapietra
Emmanuel Stefanakis
John Stell
Gilles Taladoire
Sabine Timpf
Juan Carlos Trujillo
Nectaria Tryfona
Christelle Vangenot
Peter van Oosterom
Laure Vieu
Stephan Winter
Michael F. Worboys

JoDS Editorial Board

Table of Contents

Geospatial Semantics: Why, of What, and How?

Werner Kuhn

Institute for Geoinformatics, University of Münster,
Robert-Koch-Str. 26-28, D-48151 Münster, Germany
kuhn@uni-muenster.de

Abstract. Why are notions like semantics and ontologies suddenly getting so much attention, within and outside geospatial information communities? The main reason lies in the componentization of Geographic Information Systems (GIS) into services, which are supposed to interoperate within and across these communities. Consequently, I look at geospatial semantics in the context of semantic interoperability. The paper clarifies the relevant notion of semantics and shows what parts of geospatial information need to receive semantic specifications in order to achieve interoperability. No attempt at a survey of approaches to provide semantics is made, but a framework for solving interoperability problems is proposed in the form of semantic reference systems. Particular emphasis is put on the need and possible ways to ground geospatial semantics in physical processes and measurements.

1   Introduction: Why Semantics?

In some sense, Geographic Information Systems (GIS) have always been based on semantics, sometimes even on explicitly defined semantics. For example, a GIS user in an environmental planning agency in Germany is likely to keep a heavy binder on her shelf. It is called the ATKIS Object Catalogue[1] and its role is to define the object classes and attributes occurring in topographic data, both syntactically and semantically. Similarly, land use and land cover databases have always been built according to some semantic classifications, such as the European CORINE standard [1]. So, what has changed, and what would it mean today for a GIS to be based on semantics?

The answer is that access to and use of geospatial information have radically changed in the past decade. Previously, the data processed by a GIS as well as its methods had resided locally and contained information that was sufficiently unambiguous in the respective information community [2]. Now, both data and methods may be retrieved and combined in an ad hoc way from anywhere in the world, escaping their local contexts. They contain attributes, data types, and operations with meanings that differ from those implied by locally-held catalogues and manuals. Since the semantics specified by these local resources is not machine-readable, it cannot be shared with other systems. Coping with this situation defines the challenges of *semantic interoperability* [3].

The notion of semantic interoperability is hard to pin down, for several reasons: it is somewhat redundant, there is no accepted formal definition, there are no bench-

---

[1] http://www.atkis.de/dstinfo/dstinfo2.dst_gliederung2?dst_ver=dst

S. Spaccapietra and E. Zimányi (Eds.): Journal on Data Semantics III, LNCS 3534, pp. 1–24, 2005.
© Springer-Verlag Berlin Heidelberg 2005

marks or commonly agreed challenges, the role of humans in the process is unclear, and the acronym inflation around the semantic web obscures rather than highlights the deeper research issues. Clearly, semantic interoperability is the only useful form of interoperability. In the real world, it is hard to imagine two agents interoperating successfully without a shared understanding of the messages they exchange. Therefore, it seems appropriate to define interoperability in a way that involves shared conceptualizations.

The following definition of interoperability that emerged from a geospatial context is often quoted (ISO TC204, document N271):

> *"The ability of systems to provide services to and accept services from other systems and to use the services so exchanged to enable them to operate effectively together."*

This definition is almost identical to the one in Wikipedia[2]. Such definitions are technical enough to be useful in systems engineering and testing. They also make clear that interoperability rests on services. But they fall short of establishing verifiable criteria. What does it mean for systems to operate together? And when can they be said to do this effectively?

A more precise definition of interoperability would require at least two steps: (1) identifying the vocabulary and syntax of service interfaces, and (2) defining interoperability mathematically. In this paper, I address the first requirement. Preliminary results of an ongoing debate[3] suggest that the theory of institutions [4, 5], building on category theory, supplies the necessary formal foundations for the second requirement.

Semantic interoperability is the technical analogue to human communication and cooperation. It hardly constitutes a research topic per se for Geographic Information Science, but serves as a technical goal justifying the formalization of semantics in GIS and providing measurable criteria of success for this undertaking. The research questions it raises range from those of ethnophysiography , which studies how people conceptualize landscape features, to questions about human cognition of geospatial information in general [6, 7], through formalization methods for geospatial concepts [8] and architectures for ontology-based GIS [9], to socio-economic aspects of spatial data infrastructures [10].

The notion of interoperability needs to be understood broadly enough, encompassing the interoperation between human beings and systems. But it should also remain precise enough, allowing for a common syntactic basis. While it is essential to consider the organizational and societal issues involved in information sharing [11], it is detrimental to overload the definition of technical interoperability with these aspects. Levels of interoperability should be defined incrementally, starting at the purely technical and proceeding through the organizational and social levels. Sooner rather than later, however, environments for semantic interoperability will have to include means for meaning negotiation and other ways of dealing with organizational and social contexts [12].

---

[2] http://en.wikipedia.org/wiki/Interoperability
[3] http://www.dagstuhl.de/04391/Materials/

The Muenster Semantic Interoperability Lab (MUSIL[4]), as well as other research groups (see, for example, [13]), have found that a *focus on actual interoperability problems* helps to sharpen the research questions around the broad theme of semantics of geospatial information. Investigating interoperability scenarios based on actual cases of using geospatial information for decision making provides measures of success to test specific semantic and technological hypotheses: a certain choice of concepts specified in an ontology, or certain elements in a service architecture should produce a difference in the degree of interoperability between some components. With a formal definition of interoperability, the difference could even be measured.

This paper shows what syntactic parts of geospatial information need to be specified semantically to support interoperability (Section 2); it classifies semantic interoperability problems and illustrates them through scenarios (Section 3); it postulates a solution framework inspired by spatial reference systems (Section 4), and concludes with a summary and an outlook on longer term research challenges (Section 5).

2   Semantics of What?

This section defines the bases for semantic interoperability research by asking "what needs to be semantically specified in order to support semantic interoperability?" It clarifies the notion of semantics and the syntax of the expressions which require semantics to achieve interoperability. The fundamental construct of a service interface is highlighted and analyzed. Since the perspective taken on semantic interoperability includes human beings as parts of interoperating systems, user interfaces are subsumed under service interfaces. Finally, the question "what is special about spatial" is revisited in the context of geospatial semantics.

2.1   Semantics

The only sensible use of the term "semantics" refers to the meaning of *expressions* in a language. Such expressions can be single symbols (the "words" of a language) or symbol combinations. As the term implies, they are used to express something, i.e., to communicate meaning. Neither concepts nor entities nor properties nor processes have semantics, but expressions in languages describing them do.

The relevant languages in an information system context express how human beings conceptualize something for the purpose of representing and manipulating it in machines. Many such languages exist and need semantics: programming languages, schema languages, query languages, interface specification languages, workflow modeling languages, user interface languages, sensor modeling languages, and others. Many of these languages allow users to define new symbols (for individuals, types, properties, relationships etc.). Additionally, application standards introduce all sorts of more or less controlled vocabularies (such as those in feature-attribute catalogues or metadata standards). Furthermore, free-form text entries in data and metadata collections open the gate to almost unlimited uses of natural language expressions. Coping with the semantics of all expressions in such languages is beyond current means.

---

[4] http://musil.uni-muenster.de

Restricting the expressions to those affecting interoperability will make the task more manageable.

Attaching meaning to language expressions is a *conceptual* phenomenon. Natural language symbols and expressions evoke concepts in human minds and are used to express those concepts. For example, the term "jaguar" can evoke a concept of an animal, car, or jet fighter in a human mind, with context usually picking out the intended interpretation and discarding the others. The concepts, in turn, are shaped by human experience with some real-world entities. Thereby, expressions come to refer to entities (as well as properties, relationships, and processes) in the world. This fundamental ternary meaning relationship between symbols, concepts, and entities is captured in the so-called semantic (or semiotic) triangle, going back at least to [14], but already implicit in Aristotle's work. The triangle exists in many versions; the one shown here considers the three relationships forming the edges as human activities (using a symbol to express a conceptualization of something in the real world, and to refer to that):

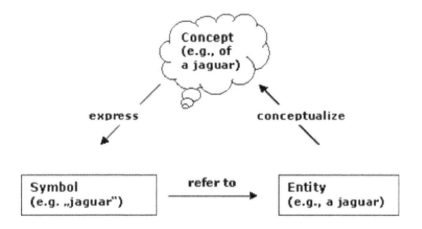

Fig. 1. The semantic triangle

The languages used in information systems are not natural languages, even if they use natural language terms. They are the results of social *agreements* in information communities on how to use certain terms; agreements which are typically more explicit than those underlying the use of natural languages. The agreements establish technical terms (say, `overlap` as a topological operator), which are recognized to have a relatively fixed meaning that is sometimes formally defined and often made explicit in the form of feature-attribute catalogues, interoperability standards, legal regulations, and other defining documents. For example, the navigation community has agreed on various forms of graph representations to model road networks for navigation purposes [15]. Codifying such agreements in ontologies is a useful first step toward semantic interoperability [16, 17].

The symbols and expressions of information system languages can be produced or consumed by machines, but acquire meaning by the same relationships as those of natural languages. The fundamental fact about meaning, that it is generated by humans and not defined by a state of the world, applies to all symbols, and independently of whether they stand for individuals (as names or constants do) or classes (as nouns or type labels do). This view of semantics avoids the pitfalls of simplistic associations between symbols and entities in the world, sometimes referred to as realist semantics [18].

Geospatial semantics, consequently, is *not* about the relationship between GIS contents and the world, and does not need to be: this relationship is already captured in the notion of *correctness* (and, more generally, integrity) of databases and information systems. Geospatial semantics is about *understanding* GIS contents, and capturing this understanding in formal theories. At the same time, one should not make simplistic assumptions about the nature of the concepts that define such understanding. They are not just individual notions, but constantly evolving and often elusive results of conceptualization processes in information communities.

Is the goal of research on geospatial semantics to fully specify the semantics of geospatial terms? Such an enterprise would be too daunting, but also unnecessary. Consider how well human communication works without precisely defined semantics. We all use one or more natural languages (such as English or Mandarin) to communicate, none of which has a formally defined semantics. Yet, we understand and cooperate with each other reasonably well, despite frequent semantic ambiguities. As human beings living in certain social contexts, we have devised means of resolving these ambiguities *as far as necessary* to make communication and cooperation successful. This fact should caution us against putting more emphasis on formalizing meaning than on the reasoning that uses these formalizations to make necessary distinctions. Nevertheless, a few words on formalization are in order before addressing the reasoning challenges posed by interoperability.

2.2   Formalizing Semantics

Since concepts (and meanings, as relationships between expressions, concepts, and the world) are not directly observable, theories of semantics have to introduce substitutes for them. They can choose to represent meaning as a relationship between symbols (symbols of a language and symbols representing concepts) or instead represent effects of meaning (for example, the actions in the world resulting from understanding an expression). The former option is taken by the field of formal semantics and constitutes the only practical approach today. The latter requires theories of action (and of the role of information in them) that are not available yet for geospatial applications. As it would compensate some shortcomings of formal semantics, I will discuss this option in some more detail in Section 4.

Formal semantics, as coming out of logic, linguistics, and computer science, establishes a mathematical basis to talk about meaning. Through model theory, it introduces the notion of possible models, formally defining the semantics of expressions [19]. These models are considered to *be* the meanings. From a conceptual point of view, they are just symbolic structures, albeit useful ones: They represent conceptualizations of entities, properties, and relationships in a domain and can therefore be tested against human intuitions about these [20]. Differences observed between the intuitions and the behavior of the models can then suggest possible changes to the models.

Thus, model theory allows, at least in principle, for empirical tests of hypotheses about the semantics of expressions. Such hypotheses are formulated in information system *ontologies*. An ontology is a "logical theory accounting for the intended meaning of a formal vocabulary" [21]. It has to be logically satisfied by its models. The closer the models correspond to the human concepts about a domain, the more useful will an ontology be. The richer the models are, the more powerful hypotheses can be tested. But the questions how detailed and how expressive ontologies should be are purely pragmatic ones. The answers depend on the levels of interoperability to be supported.

The perceived limitations of model-theoretic approaches to meaning are largely due to a limited notion of models, which are often restricted to sets. Unstructured sets are almost always too weak to serve as interesting conceptualizations of the world. For example, they cannot adequately model mereological relationships, which are essential for spatio-temporal applications [22]. Humans do not understand domains as sets of things and subsets formed by predicates, but through their behavior and the actions that can be performed in them [23]. A road is a road by virtue of linking places in a way affording cars to drive from one place to another. A lake is a lake because it holds standing water and serves as a (possibly empty or frozen) water reservoir, for swimming, sailing, and even driving [24]. Modeling such processes through operations creates an algebraic structure, which captures meaning through models and associated mappings (morphisms) within and across domains [25].

A more fundamental pitfall of model theory (and of any other theories of meaning based on symbolic structures) lies in the *symbol grounding* problem [26]: how do the language (and model) symbols acquire meaning? Describing their meaning by other symbols begs this question. Languages are much richer systems than formal symbol systems. The meaning of their expressions emerges, lives, and evolves in language users and communities, where human bodies and minds communicate [27]. Meanings are not fixed and cannot be assigned to symbols independently of how these are used. All symbolic approaches to semantics, therefore, are necessarily limited in scope and need to be complemented by studies of language use and evolution [28]. Breaking out of the symbolic cage will eventually require pursuing the option of accounting for meaning by modeling observable effects in the world (see Section 4).

2.3   Semantics of Services

The focus on semantic interoperability picks out a small subset of languages and defines their semantics: those used to specify and invoke services. Consequently, the semantics required to achieve interoperability is that of expressions built from symbols in service descriptions. In the semantic web context, various languages are used to write such expressions. For example, WSDL (the Web Service Description Language) allows for syntactic descriptions of web service interfaces and OWL-S (the service ontology of the Web Ontology Language) has been proposed for semantic specifications of services. More comprehensive service modeling efforts like WSMF (the Web Service Modeling Framework) are under way [29].

It is not always clear how service descriptions are to obtain semantics and what additional modeling languages may be needed. In particular, it is difficult to say something meaningful and useful about the operations performed by services. The semantic web approach is to specify input and output types, pre- and post-conditions, and taxonomies of service types. But the form and use of pre- and post-conditions are unclear, the granularity of service taxonomies is too coarse, the service types themselves are not semantically defined, and the algebraic structure that operations impose on domains cannot be captured. We simply do not know yet what needs to be said, and how, about the semantics of services to make them semantically interoperable. It makes sense, therefore, to take a step back and study their vocabulary and syntax in more detail.

2.4 Interfaces

Agents, computational and human, interoperate through interfaces. For meaningful and useful interoperation, these interfaces need well-defined semantics. In today's GIS service architectures, the interfaces are those of web services, interacting with each other or with human users. The success of the transition from the distributed computing platforms of the nineties and earlier architectures to web-based service architectures depends to a large extent on the development of techniques to specify and query the semantics of service interfaces.

The idea of an interface is well understood in computer science and provides an excellent basis for modeling its semantics. For more than thirty years, software engineers have known that the semantics of data and operations are inseparable and that they are best explained in terms of interfaces of software components [30]. Based on this insight, the Open Geospatial Consortium (OGC[5]) has transformed the architecture of GIS over the past decade. Acknowledging the role of interfaces for distributed computing [31] in the GIS area, OGC identified the interface of software components as the key ingredient of GIS technology that needed standardization in order to achieve interoperability [32]. The result is a series of syntactic interface specifications, establishing protocols for components exchanging geospatial information. The information so exchanged can contain features, maps, coverages, or metadata.

These service interface standards establish syntactic interfaces and protocols for invoking system behavior, but do not specify the intended meaning of their terms in machine-readable form. Consequently, individual components can only be tested for conformance to specifications, but not for interoperability with each other. The need to attach semantics to the syntax specifications has been well recognized in OGC from its early days. It led to the vision of services interoperating within and across information communities through semantic translation [33].

From a semantics point of view, one wishes that OGC's focus on interfaces had not been diluted by the recent shift of attention to data exchange through GML (the Geography Markup Language). GML is a very useful and necessary schema language, but major efforts are now going into the exposure and harmonization of database schemata, activities which used to be considered unnecessary and even dangerous (because they lock providers and users into fixed data models). To bridge between the original vision and the current practice, the oxymoron of "data interoperability" has crept into industry and agency jargon. It suggests, figuratively speaking, that the flour and eggs in your kitchen interoperate among themselves to prepare pancakes.

---

[5] http://www.opengeospatial.org

The point is not that humans need to be involved (this may or may not be the case), but that it takes *operations* to interoperate, not just data [34].

The *user interfaces* of software systems share many properties of component interfaces: they contain commands with well-defined syntax and observable semantic properties. Input parameters are being set, operations get executed, and results are returned. Thus, user interfaces exhibit, at least in principle, the same syntactic structure and behavior as internal component interfaces. Their input and output types are often more complex, but form interfaces of the same kind as those of object classes. These interfaces are equally or more important in the quest for semantic interoperability than internal interfaces, because, ultimately, information is always from and for human beings. Users are essential parts of interoperating systems.

2.5 Signatures

The interface of a service is formally captured by its signature. A signature describes a service's type information, consisting of the input types, output types and names of the operations offered by the service. Without loss of generality, I assume here that a service consists of a single operation. A core geospatial example of a service is the specification of the distance operation in the ISO Spatial Schema standard [35]:

```
GM_Object:: distance (geometry: GM_Object): Distance
```

This signature says that the distance operation shall be applied to a geometric object (GM_Object), takes as input another such object (called geometry), and returns a value of type Distance. The two geometries could be points or other geometric objects.

The notion of a signature is fundamental to algebra and category theory [36] and plays a central role in algebraic software specifications [37, 38]. Signatures contain symbols expressing

- values
- objects
- functions.

All these symbols stand for either an individual or a type. In the distance example, GM_Object expresses a type of object, distance an individual function, geometry an individual object, Distance a type of value. A call to a service implementing the distance operation would return an individual value. Note that the semantic characteristic of a measuring unit is normally considered to be a part of the function representing the measurement. Strictly speaking, an individual function implies a particular choice of unit (such as meters). For example, the interpretation of a result value 100 as 100 Meters would be given by the particular distance function.

Taken together, the type symbols in the signature describe the type of the distance function. The standard mathematical form of the signature shows this more clearly:

```
distance : GM_Object x GM_Object → Distance
```

It says that distance is a function of type $GM\_Object^2 \rightarrow Distance$ and treats the two geometries symmetrically, as one would expect from a distance function. A shortcoming of today's service specification and implementation languages is that they rarely allow for this decoupling of functions from single value or object types.

Thus, service signatures as a whole, and the symbols they contain, express conceptualizations of a domain in terms of values, objects, and functions. Functions can have any number of arguments, including none, in which case they are individual values (such as `True` and `False`) or objects. The following table summarizes the kinds of concepts expressed by service signatures, with each cell showing an example from the distance service:

Table 1. Kinds of concepts expressed in service signatures

Concept	**Value**	**Object**	**Function**
Individual	`100`	`Point_A`	`distance`
Type	`Distance`	`GM_Object`	`GM_Object`$^2 \rightarrow$`Distance`

The different kinds of concepts are closely related to each other in two orthogonal ways:

- each individual (value, object or function) is an instance of a type;
- values and objects are the arguments and results of functions.

Our analysis of service interfaces has thus revealed a well-defined and relatively small set of semantic elements and relationships that need to be defined to enable semantic interoperability.

2.6 Geospatial Semantics

In the absence of a general theory of service semantics, it is hard to state clearly why and how geospatial services may be special. At the structural level of establishing semantics for service signatures, there does not seem to be anything special about space (or time). Yet, the geospatial data types and operations occurring in these signatures, and the conceptualizations underlying them, are characterized by some important properties (see also [39] for an implementation-oriented view):

1. Geospatial data and services contain symbols whose meaning is not only a matter of convention, but grounded in physical reality. For example, a wind direction returned by a weather service or a water level measured by a gauge have an *observable grounding in the world*. Conversely, the meaning of their measuring units, of a currency amount, or of a single-click purchase at an e-commerce site is purely conventional. Because of this physical grounding of some concepts, explaining the semantics of geospatial information requires *measurement ontologies* [40] that are tied to existing standards in science and engineering [41].
2. At the same time, geospatial information is often based on *human perception and social agreements*, combining objective measurements with subjective judgments. Coping with the meaning of qualitative judgments (say, of statements on landscape aesthetics) or of social constructions (like neighborhood classifications), and providing mappings among them [42], are probably the biggest challenges ahead to make geospatial information more meaningful and shareable. They require a layered architecture of ontologies [43], not just different unconnected perspectives or different levels of application specificity.

3. A special case of social agreements are *geographic names* and other identifiers of geospatial entities. Geographic name registries in the form of gazetteers will need better translation and geo-referencing capabilities. Object identifiers in different databases across information communities will need to be linked. For example, the same petrologic sample may be registered under different identifiers and referenced to different geographical names in various online databases supporting geochemical analyses [44].

4. Space and time are primarily understood through *processes*: we locate stuff because we can move it, we use distances and directions to navigate, and we determine when to leave the beach by estimating the speed of an advancing storm. This process-nature of geospatial information challenges the entity-bias of the semantic web and geospatial data models [45], though the challenge as such is surely not unique to geospatial applications.

5. Geospatial ontologies can be seen as "GIS at the type level". They should provide reasoning capabilities (spatial and non-spatial) about *types* of geospatial values, objects, and functions, similar to the reasoning provided by GIS about their instances. For example, to determine the feature types to be retrieved for flood risk assessment, one has to reason about spatial relations like proximity between entities and rivers, independently of particular feature instances.

6. *Vagueness, uncertainty,* and *levels of granularity* are fundamental to geospatial information. Theories of vagueness and uncertainty, as well as mappings among spatio-temporal [46, 47] and semantic [48] granularity levels of ontologies are therefore essential ingredients of geospatial semantic theories.

Clearly, this is an open-ended list of characteristics, and none of them applies only to geospatial information. But it is useful to keep such considerations in mind when evaluating approaches to semantic modeling for geospatial domains. Equally important, however, is to clarify the interoperability problems to be solved through the semantic models. These problems are the subject of the next section.

3   A Classification of Semantic Interoperability Problems

The premise that interoperability is the technical reason to model the semantics of geospatial information, together with the defining role of services for interoperability, allows for identifying and classifying interoperability problems. This section introduces three problem classes through geospatial examples. The problem classes are orthogonal and complementary to the types of semantic heterogeneities (naming, conceptual) discussed in the interoperability literature [49]. They capture specific reasoning challenges that arise in the course of making systems and services interoperable. The necessary reasoning is often referred to as matchmaking and is here briefly introduced before discussing the problem classes.

3.1   Matchmaking for Interoperability

Matchmaking is the fundamental procedure enabling semantic interoperability [50]. It is a *reasoning* process with the goal of deciding whether an information offer matches a request. The reasoning can be performed by humans or software or a combination of both. Its result can be binary (match or not) or a measure for the degree of match, i.e., for similarity.

The main task in matchmaking is to determine and resolve semantic *heterogeneities* between requests and offers. There are *naming* heterogeneities (different expressions for the same concept) or *conceptual* (a.k.a. cognitive) heterogeneities (different concepts expressed by the same symbols). The naming heterogeneities are sometimes further subdivided into syntactic (different symbols) and structural (different expressions). An example for a syntactic naming heterogeneity is a distance value expressed as a floating point number or as a distance type. A structural naming heterogeneity is that between a location expressed by two separate coordinates or by a point data type. A conceptual heterogeneity would be that between a distance computed on the sphere or in a plane.

Clearly, matching data to specifications and resolving the corresponding heterogeneities is much easier than matching services to specifications. These two cases define the first two classes of interoperability problems. An even more difficult case of matchmaking is the reasoning to determine whether and how services can be composed to produce a desired behavior. The matches sought are then between the services (to form a combined service) and between the composed service and the request. This case defines the third problem class. Each of the three problem classes includes the previous one as a part of the problem.

3.2 Data Discovery and Evaluation

Today, the bulk of digital geospatial information resides in databases and files. Users of these data need information on what they mean. No matter whether they access the data through web sites, database queries, import functions, connections to data warehouses, or OGC web services – at some point they will receive values, attribute names, and complex objects. Searching for data sources and evaluating their contents define the first class of semantic interoperability problems.

Consider a hydrologist searching for information on water levels of the river Elbe [51]. She may be in charge of issuing flood warnings or monitoring ecological indicators. Among the data sources at her disposal are gauge readings from different stations. Three examples of water level data providers on the World-Wide Web are:

- The German Federal Agency for Hydrology[6];
- The German Electronic Information System for Waterways[7];
- The Czech Hydrometeorological Institute[8].

The data offered by these sources consist of attributes for station names and water levels, time stamps, station locations, river names, and additional hydrological information on water discharge and the like.

Interfaces to data represent the special case of (service) interfaces without computational functionality. Their structure can therefore be described by signatures, and the

---

[6] http://www.bafg.de/php/elbe.htm
[7] http://www.elwis.de/gewaesserkunde/Wasserstaende/Wasserstaende_start.php? target = 2 & gw=ELBE
[8] http://www.chmi.cz/hydro/SRCZ04.html

concepts expressed by the symbols are a subset of those in Table 1, leaving away its last column:

Table 2. Kinds of concepts symbolized in data repositories

Concept	**Value**	**Object**
Individual	`158`	`Elbe`
Type	`Höhe`	`WasserstandMessung`

For example, the data source providing the value of `158` in Table 2 declares it to be of type `Wasserstand` for a given `Pegel` (station), `Datum` (date) and `Uhrzeit` (time). Even if the German terms could be interpreted by a client (human or software), the measuring unit, reference level, and measurement or averaging process for the water level remain unspecified.

The matchmaking needed to discover and evaluate data sources has to resolve such ambiguities. Existing metadata standards and catalog services do not support this process well [51]. Their contents and search procedures are keyword-based, similar or inferior to those of internet search engines, with no way of resolving naming or conceptual heterogeneities. The keywords are not treated as values of different types, and normally not taken from controlled vocabularies. They are just strings, to which machines cannot attach any meaning, and humans may or may not apply the right interpretation. So far, the main efforts in using the semantic web for geospatial applications have been geared to improve this situation [52].

3.3   Service Discovery and Evaluation

While only a relatively small amount of geospatial information is provided in service form today, the number and computational power of geospatial information services is growing rapidly. In addition to data access, such services offer processing and portrayal capabilities. They may be coupled to specific data sources or applicable to data from multiple sources. Discovering and evaluating such services represents the second class of interoperability problems.

The additional (and more challenging) semantic issues in this second problem class arise from the need to reason about the *functionality* of services. Describing the meaning of an operation like `distance` is far from trivial: the operation signature can refer to many different kinds of distances (metrics), ranging from the path length in a graph through the Euclidean or Manhattan distances in the plane to a geodesic or straight-line distance on the surface or across the earth [53]. All of these distance operations have the same signature shown above (i.e., they are of the same type).

Obviously, the functionalities of more complex geoprocessing operations (such as buffering or overlay, but also topological operators [54]) pose even harder semantic challenges. If the functionality descriptions become too complex, they are unlikely to be produced by service providers and understood by clients. But if they are too simple, recall and precision in discovery and evaluation are reduced. Furthermore, the descriptions need to support the reasoning necessary to match service offers to requests. If this reasoning becomes too expensive, it threatens the efficiency of service discovery and evaluation. Traditional specification methods from software engineering turn out to be either too weak in expressiveness or too complex for the available reasoning mechanisms [40]. Process ontologies seem a promising alternative [55], but their contents and associated reasoning methods are not yet clear, and they lack spatio-temporal notions [47].

The semantic heterogeneities in this second class of semantic interoperability problems concern all six kinds of symbols shown in Table 1. But peculiar to this problem class are the semantics of *function* types and individuals. For example, the type $GM\_Object^2 \rightarrow Distance$ of a distance function needs an interpretation and so does the individual function symbol $distance$. The function type specified in the ISO Spatial Schema Standard is quite precise, with an explicit result type $Distance$. However, implementations typically use a more generic result type, such as a floating point number. The semantics of the function type (e.g., $GM\_Object^2 \rightarrow Float$) then becomes highly ambiguous and can be interpreted as any real-valued property of two geometric objects. As a consequence, the full semantic burden rests on specifying the individual function identifier, $distance$.

Some semantic heterogeneities in service signatures can be resolved through spatial reference systems [56]. These provide information about the spaces in which the arguments of an operation (the geometric objects) are embedded. For example, the spatial reference system of two geometric objects going into a distance operation may be a plane coordinate system tied to a certain map projection. One can safely assume that a distance should be calculated in that same reference system, though this still leaves open which metric it uses. Also, a distance service that is decoupled from a data source would either need to be restricted to a fixed reference system and metric, with service metadata describing these choices, or carry the generic and complex functionality for all possible combinations. Finally, the case of the two geometric objects having different reference systems needs to be resolved (as, for example, in [35]: "If necessary, the second geometric object shall be transformed into the same coordinate reference system as the first before the distance is calculated").

Current GIS practice does not suffer much from this second class of interoperability problems. It uses coarse-grained generic service interfaces, like those of feature servers, and combines them with database schema exchange through GML. Feature, coverage, and map services as specified by OGC essentially provide semantics-neutral wrappings for repositories of vector, raster, and map data. This puts us back to problem class one. Admittedly, the idea of finer-grained service interfaces, which used to be seen as the core of interoperability in the geospatial area, has been hampered by complexities like those exhibited in CORBA applications [57]. But by breaching the information-hiding principle of object orientation and exposing internal data formats, data get separated from the operations they were created for (or by), and a heavier price has to be paid to restore meaning to them. The general evolution toward finer-grained functionality offered over the web may bring the second class of interoperability problems to the forefront again.

3.4 Service Composition

Full-fledged semantic interoperability involves not just individual services to be discovered and used, but multiple services interoperating with each other. The third semantic interoperability problem class is defined by the semantic issues raised through automated or manual compositions of services to produce more complex services or entire applications. It is characterized by the need for these services and their clients to share an understanding of what the services do and what goes across their interfaces [58].

Consider a service to compute the outline of a toxic cloud at some point in time after a chemical accident, taking as inputs a report on the chemical accident and data from a weather service [59]. Assume that the accident report provides location, time, type and emission rate of chemical, while the weather service returns wind direction and speed. The values, objects, and functions involved pose the same kinds of semantic heterogeneity issues as in problem class one. For example, the functionality of the plume calculation service uses a certain spatio-temporal resolution and a threshold concentration of the chemical to determine the outline.

Matchmaking in this problem class, however, is more complex. Service requests may depend on previously found offers for other services. This interdependency leads to a more involved reasoning process, spanning over requests and offers of multiple services. For example, the wind information supplied by the weather service may differ from the one expected by the plume service (e.g., it may follow the meteorological standard of westerly wind blowing *from* the west, while the plume service might expect a vector direction, such that a 270° wind blows *to* the west). The request for weather information, then, depends on the plume calculation.

Furthermore, this problem class involves *mediation* between the output provided by one or more services and the inputs required by another. For example, a semantic translation from one conceptualization of wind direction to the other may be needed. This translation task remains the biggest challenge of semantic interoperability, particularly when it concerns service functionality.

4   A Framework for Solving Semantic Interoperability Problems

What methodological approach is required to solve the semantic interoperability problems defined in the previous section? All three problem classes have been characterized as involving matchmaking, i.e., reasoning about the compatibility of offers and requests for data or services. This reasoning perspective emphasizes the need for approaches that go beyond the construction of ontologies and involve their use for discovering, evaluating, and combining geospatial information. Semantics-based GIS are about reasoning, not just about ontologies. This section presents some thoughts on the reasoning requirements and a methodological framework in the form of semantic reference systems.

4.1   A Geospatial Analogy

One can think of ontologies as establishing conceptual *"coordinate systems"*, i.e., frames of reference for positioning concepts in a certain context. For example, the concept of a car can be specified in an ontology as a specialization of a vehicle. If vehicles are in turn specializations of conveyances, one can conclude that cars are special conveyances, sharing all their properties and relationships. This taxonomic reasoning is fundamental to most ontology applications today. It has a spatial analogue in (set) containment: all instances of cars are contained in the set of all conveyances.

Taxonomic reasoning is useful, but insufficient for the matchmaking tasks described above. Equally or more important are *non-taxonomic relationships*, e.g., that wind direction and speed are parts of a wind force, or that a car can move on roads. Reasoning with these is much harder, as it is not of the simple set inclusion kind required for taxonomies, but depends on the semantics of each relationship. For example, complex relationships between moving air masses, locations, and measurement scales define a concept like wind direction [59].

Coordinate systems in geometric spaces allow for computing distances. Conceptual coordinate systems should support the computation of conceptual distances and *similarities* based on these. Similarity theories exist, also for geospatial concepts [60], and show the importance of capturing the context-dependence of human similarity judgments. Several approaches exist to cope with this problem. They are, for example, based on modeling the use of entities [61] or on "factoring out" context through relative similarities (a is more similar to b than to c, or a is more similar to b than c is to d). In addition to context, all similarity theories are challenged by the question whether they should compare individuals, types, individuals to prototypes, or any combinations of these.

Context is an overloaded term and has many aspects. Some of them are relatively easy to handle through domain separation (e.g., the difference between banks in a financial and in an ecological context). Others are much harder to deal with, because they result from different *groundings* for the meaning of a symbol. For example, the difference between a mathematical (blowing to) and meteorological (blowing from) conceptualization of wind direction has a physical grounding. As long as such groundings are not represented in ontologies, no amount of taxonomic, non-taxonomic, or similarity reasoning can distinguish them or even reconcile their differences. Today's ontologies are islands in a sea of different conceptualizations, which are hard to connect [62]. They lack the means to ground conceptualizations in reality and therefore cannot solve the symbol grounding problem [26]: they do not anchor their conceptualizations in reality.

Geometric coordinate systems, by contrast, are anchored in physical features, such as fundamental geodetic points (materialized in monuments) and the rotation axis and parameters of the earth. This anchoring is called a geodetic datum [56]. The use of any coordinate system, spatial or otherwise, without anchors in reality is limited to local reasoning and cannot explain how the "coordinated" ideas relate to the world. Due to its lack of grounding, ontological reasoning today derives mostly local containment relations between the extensions of concept specifications. These are neither invariant nor translatable across multiple ontologies.

This rather loose analogy between meaning and geometry can be made stronger in several ways. Fabrikant, for example, is exploiting it for information access and visualization [63]. Gärdenfors has taken the idea of representing concepts geometrically very literally, in his notion of conceptual spaces [18]. His theory provides a solid mathematical basis for the analogy between concepts and geometric spaces and exploits it fruitfully for all sorts of representation and reasoning challenges, in particular for similarity measurements and transformations. Related in spirit, but with more emphasis on computational processes and less detail on representations, I have proposed the notion of *semantic reference systems* [64]. It takes the spatial analogy seriously in terms of the reasoning requirements it implies, and derives these from the computational power afforded by spatial reference systems.

4.2 Semantic Reference Systems

The information provided by a GIS is only useful if it rests on a well-defined spatial reference system. For example, way-finding directions refer to landmarks in reality and use distance and direction measurements anchored in physics. Maps represent the territory in a certain map projection, which allows for calculating distances and directions. Latitudes and longitudes can be traced back to arbitrarily exact locations on the surface of the earth. The meaning of a coordinate in a GIS database is entirely specified through the associated spatial reference system, and the meaning of geometric computations (such as distances) can be tied to the same system. The same is true for temporal data and reference systems. In other words, for coordinates and time stamps, we already have theories "accounting for the intended meaning of a formal vocabulary" [21], though they are based on algebra rather than logic.

Would it not be nice to have an equally powerful method of disambiguating the meaning of the remaining symbols carrying geospatial information, such as "wind direction" or "water level"? As Chrisman has already suggested [65], users of geospatial information should be able to refer thematic data to attribute reference systems, just as they refer geometric data to spatial reference systems. This idea suggests that the symmetry between the two components of Goodchild's geographic reality (a spatio-temporal location vector and an attribute vector [66]) is incomplete without reference systems for the attributes.

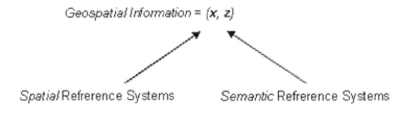

Fig. 2. Reference systems for interpreting geospatial information

From a semantic interoperability perspective, this requirement extends beyond attributes to cover all parts of service interfaces introduced in Section 2. I have therefore called for *semantic reference systems* to offer the necessary representations and reasoning capabilities for

- *referencing* symbols to concept specifications
- *grounding* concept specifications in physical reality
- *projections* among the semantic spaces
- *transformations* among different semantic reference systems.

While referencing is handled by ontologies today, grounding is not, projections are limited to piecewise generalizations in taxonomies, and translations need to make strong assumptions about shared conceptualizations. If computational solutions to these challenges appear unlikely, imagine the situation for theories of space as it was before Descartes invented coordinate systems. Sharing information about location, shape, and extent of things in the world was then probably just as difficult as sharing of non-geometric information is today. Fortunately, a solid mathematical theory of coordinate systems and their physical grounding is now available and can serve as a model for semantic theories and the capabilities they need to offer. Indeed, the best way to look at spatial reference systems is as the special (geometric) cases of semantic reference systems.

The representational, computational [67] and institutional challenges posed by this vision are substantial. However, they will need to be met, if the promise of semantic interoperability through the semantic web is to be fulfilled: a web "...in which information is given well-defined meaning, better enabling computers and people to work in cooperation" [68]. A formalization of Gärdenfors' conceptual spaces is likely to play a key role in implementing the reasoning capabilities, particularly projections and transformations [69].

Since the toughest challenge lies in the need for grounding, the rest of this section sketches two promising directions to pursue for grounding geospatial information in reality: based on image schemas and on measurements. The two ideas are connected through the key insight that image schemas are abstractions from experiences with processes in the world, which in turn have observable effects. Their emphasis on the links between processes, observations, and information is characteristic for a range of recent work related to geospatial information (see, for example, [8, 70, 71]).

4.3 Grounding in Image Schemas

Grounding the meaning of symbols through symbols is an oxymoron. Yet, some symbols (or symbolic structures) are more easily grounded than others. In other words, human interpretations of them are more likely to agree across domains and cultures through a shared understanding rooted in some physical processes. The claim behind the idea of an image-schematic grounding of ontologies is that symbolic structures representing image schemas possess this property.

Image schemas, as introduced by cognitive linguists and philosophers like Len Talmy, Ron Langacker, George Lakoff, and Mark Johnson (see [72] for a recent survey) are mental patterns shaping our thought, action, and language. They are rooted in our *bodily and cultural experiences* and extract the common structure of these. For example, the container image schema abstracts from our experience of dealing with and reasoning about anything that can contain anything else, such as cups, human bodies, or rooms. This experience is characterized by the processes of putting something into a container, discovering that it is inside, and taking it out again. Similarly, we build patterns from our experiences with surfaces, paths, links, covers, parts and wholes, centers and peripheries, force feedback, and an open-ended series of other structures (see p. 126 of [73] for a list).

As the examples show, many image schemas are spatial (mostly topological), and our understanding of them is process-driven, with an *algebraic structure* linking the processes (e.g., what is taken out of a container has been put into it before). The spatio-temporal nature and process character of image schemas, together with their claimed universality across languages and their cognitively fundamental role, predispose them as candidates for grounding the meaning of symbols in experiences of physical reality. Furthermore, image schemas are typically combined to generate more complex patterns, and transformed to emphasize certain parts. Finally, they have long been suspected to define those relationships that remain invariant under all sorts of semantic mappings, such as metaphor, translation, and conceptual blending [74].

For a geospatial semantics example, consider how to capture the concept of a road. One can do this in conventional taxonomic fashion by sub-classing it from a concept like construction (which in turn is sub-classed from artifact, physical object, etc.). With sophisticated ontologies (such as DOLCE[9]), one can even add constraints on a driving activity involving roads and vehicles. But this approach assumes that the upper levels of these ontologies are unambiguously understandable across domains. It also lacks the expressiveness to differentiate multiple ways for an entity (such as vehicle vs. a road) to participate in a process (such as driving).

Alternatively, one can introduce upper levels that represent image-schematic concepts, such as paths, containers, and surfaces. A combination of paths with containers, such that a vehicle is a container moving on a path (i.e., acting as a conveyance), results in rich semantics for all participating concepts (vehicle, road, driving), while keeping the concept hierarchies flat. This shows that image schemas capture essential behavior of entities and provide useful grounding in our physical experience (of moving and containing, in this case). However, the question remains of *how* an image schema should be represented in order to evoke the intended interpretations. Still, in a sophisticated model theoretic view (see Section 2), this question can be answered empirically. Such ideas are currently pursued in the SeReS project[10].

4.4 Grounding in Measurements

Geospatial information serves to understand the human environment and to decide on actions in it. One of its most important sources will soon be sensor networks. Despite the fact that many GIS applications produce and use information that remains valid for some time (such as geological or land cover maps), more and more decisions in organizations and societies rely on timely observations of the environment, and often almost real-time data (e.g., about traffic conditions). The ground-breaking technology of *sensor networks* allows for supporting such decisions in entirely novel ways. For example, vehicle navigation systems can access sensor data from stations or from vehicles ahead and alert drivers of conditions regarding weather, congestions, accidents, construction sites, and the like. Similarly, decisions on human activities affecting the environment will be supported by more current, higher resolution, and more relevant environmental data.

A quantum leap for information processing and interoperability will result from the possibility to insert geospatial information into *feedback loops*, where an activity is guided by observations on the effects of previous actions. For example, water level

---

[9] http://www.loa-cnr.it/DOLCE.html

[10] http://musil.uni-muenster.de/index.php?m1=Research&m2=Semantic&m3=Summary

readings from a network of gauges in a river basin could be combined with hydro-logical models to guide preventive and corrective action in a flooding situation, and the results of these actions will become directly observable through the same network. International regulations (such as the European Water Framework Directive[11] or the INSPIRE[12] project targeting a European environmental information infrastructure) are now creating government mandates to collect and disseminate such information. Similar regulations can be expected in the security and health sectors.

These exciting technological and social developments add further semantic chal-lenges to cope with, but also suggest a novel approach to grounding: All information ultimately rests on observations, whose semantics is physically grounded in processes and mathematically well understood [75]. Exploiting this foundation to understand the semantics of information derived from observations would produce more powerful semantic models. For example, a service interface consuming weather information could refer to a standard library of meteorological measurement types with well-defined semantics.

The Open Geospatial Consortium (OGC) has recognized the huge potential of these developments and created the Sensor Web Enablement initiative (SWE[13]). It proposes the special feature type of a "sensor observation" for comprehensive meas-urement data. Measuring units are to be dealt with through reference systems, which are nuclei (and precursors) for semantic reference systems. Their role is to define the context for interpreting measured values and to constrain the valid operations on them. This is, of course, a modern technological manifestation of Stevens' theory of measurement scales [75].

5   Conclusions and Outlook

In this paper, I have looked at geospatial semantics from the perspective of semantic interoperability. I argued that interoperability is the *raison d'être* for semantics re-search, and that it makes problems and hypotheses more specific and easier to test than a general "semantic studies" approach to geospatial information. I have treated semantics as a conceptual phenomenon, involving language expressions and human concepts, rather than as a correspondence between terms and situations in the world. But I have also stressed the need to anchor concept specifications in reality.

Against this background, I have asked what needs to be semantically defined in or-der to support semantic interoperability. The answer was that it is expressions built from service signatures, which can be considered the syntax of interoperability. Three classes of semantic interoperability problems were defined and discussed with respect to their reasoning challenges: data discovery and evaluation, service discovery and evaluation, and service composition. State-of-the-art solutions address mainly the first class of problems, while service semantics remains elusive, both for discovery and composition. Thus, I concluded that more powerful techniques than today's ontolo-gies and reasoning environments are required to support semantic interoperability.

---

[11] http://europa.eu.int/comm/environment/water/water-framework/index_en.html
[12] http://inspire.jrc.it/home.html
[13] http://www.opengeospatial.org/functional/?page=swe

To serve this goal, I presented a framework for referencing, grounding, and mapping geospatial information in the form of semantic reference systems. The idea of such systems has been inspired by spatial reference systems and is intended to lead to analogue capabilities for non-coordinate symbols. Referencing is mostly addressed through the work on geospatial ontologies today. Grounding has been found to be particularly relevant for geospatial applications. As it has received little attention in the ontology literature so far, I sketched two complementary approaches to it: anchoring concept specifications in image schemas and in measurements. Mappings, in the form of projections and transformations between ontological specifications, will require such grounding and present the next major research frontier.

A theory of semantic translation, capable of mapping geospatial information within and across the boundaries of information communities, should indeed be seen as the overall goal of research on geospatial semantics. In the spirit of the geometric analogies used here and elsewhere [18], one can speculate that such a theory might take the form of an *"Erlangen program" of meaning*: a formalization of semantics based on invariants under certain groups of transformations, very similar to Klein's seminal work that put geometry on modern mathematical foundations (and to some extent helped create these foundations) in the second half of the 19th century [76]. To connect such a theory to reality, the meaning of symbols used in geospatial information will eventually need to be tied to an understanding of how information supports decisions on actions, and how observations of the effects of actions in turn generate new information.

The big practical challenges ahead lie in an evolution from semantic interoperability to *semantic integration* of geospatial information. All interoperability problems are also integration problems. In order for two system components to interoperate, they must share an integrated view of some information contents. Information integration, however, goes far beyond interoperability and includes issues like question answering with multiple information sources of different quality, meaning negotiation, or knowledge management in large organizations. Starting with a focus on semantic interoperability makes the posing and solving of research problems more manageable, but the larger perspective on information integration already needs to guide our methodology.

Acknowledgments

Discussions with the members of MUSIL (http://musil.uni-muenster.de) and with too many colleagues to be named have greatly influenced and improved the ideas presented here. Funding from the University of Münster, the European ACE-GIS project (IST-2002-37724) and the *meanings* project in the German Ministry of Science *Geotechnologies* program is gratefully acknowledged.

References

1. EEA, *CORINE Land Cover (Technical guide)*. European Environmental Agency, Commission of the European Community, 2000.
2. Bishr, Y., et al., *Probing the Concept of Information Communities - A First Step Toward Semantic Interoperability*, in *Interoperating Geographic Information Systems (Proceedings of Interop'97)*, M.F. Goodchild, et al., Editors. 1999, Kluwer: pp. 55-71.

3. Sheth, A.P., *Changing Focus on Interoperability in Information Systems: From System, Syntax, Structure to Semantics*, in *Interoperating Geographic Information Systems*, M.F. Goodchild, et al., Editors. 1999, Kluwer: pp. 5-30.

4. Goguen, J.A. and R.M. Burstall, *Institutions: abstract model theory for specification and programming*. J. ACM, 1992. 39(1): pp. 95-146.

5. Goguen, J., *Information Integration in Institutions (draft)*, in *Memorial volume for Jon Barwise*, L. Moss, Editor. (to appear). http://www.cs.ucsd.edu/users/goguen/pps/ifi04.pdf.

6. Montello, D. and S. Freundschuh, *Cognition of Geographic Information*, in *A research agenda for geographic information science*, R. McMaster and E. Usery, Editors. 2005, CRC Press: pp. 61-91.

7. Mark, D.M. and A.G. Turk. *Landscape Categories in Yindjibarndi: Ontology, Environment, and Language*. in *Spatial Information Theory - Foundations of Geographic Information Science, COSIT 2003, Kartause Ittingen, Switzerland*. 2003: Springer, Lecture Notes in Computer Science 2825: pp. 31-49.

8. Kuhn, W. *Modeling the Semantics of Geographic Categories through Conceptual Integration*. in *Geographic Information Science - Second International Conference, GIScience 2002, Boulder, CO, USA, September 2002*. 2002: Springer, Lecture Notes in Computer Science 2478: pp. 108-118.

9. Fonseca, F.T., et al., *Using Ontologies for Integrated Geographic Information Systems*. Transactions in GIS, 2002. 6(3): pp. 231-257.

10. Onsrud, H., et al., *The Future of the Spatial Information Infrastructure*, in *A Research Agenda for Geographic Information Science*, R.B. McMaster and E.L. Usery, Editors. 2005, CRC Press: pp. 225-255.

11. Harvey, F., et al., *Semantic Interoperability: A Central Issue for Sharing Geographic Information*. Annals of Regional Science, 1999. 33 (2)(Geo-spatial data sharing and standardization): pp. 213-232.

12. MacEachren, A.M., M. Gahegan, and W. Pike, *Visualization for constructing and sharing geo-scientific concepts*. PNAS, 2004. 101: Mapping Knowledge Domains: pp. 5279-5286.

13. Visser, U. and H. Stuckenschmidt. *Interoperability in GIS - Enabling Technologies*. in *5th AGILE Conference on Geographic Information Science*. 2002. Palma de Mallorca, Spain: pp. 291-297.

14. Ogden, C.K. and I.A. Richards, *The Meaning of Meaning*. 1946: Harcourt, Brace & World.

15. Timpf, S., *Ontologies of Wayfinding*. Networks and Spatial Economics, 2002(2): pp. 9-33.

16. Winter, S. and S. Nittel, *Formal information modelling for standardisation in the spatial domain*. International Journal of Geographical Information Science, 2003. 17(8): pp. 721-742.

17. Kuhn, W., *Ontologies in support of activities in geographical space*. International Journal of Geographical Information Science, 2001. 15(7): pp. 613-631.

18. Gärdenfors, P., *Conceptual Spaces - The Geometry of Thought*. 2000: Bradford Books, MIT Press.

19. Hodges, W., *Model Theory*. 1993: Cambridge University Press.

20. Grüninger, M. *Model-theoretic approaches to semantic integration (extended abstract)*. in *Dagstuhl Seminar on Semantic Interoperability and Integration*. 2004. Dagstuhl, Germany. http://www.dagstuhl.de/files/Proceedings/04/04391/04391.GruningerMichael3.Paper!.pdf.

21. 21. Guarino, N., *Formal Ontology and Information Systems*, in *Proc., 1st Int. Conf. on Formal Ontology in Information Systems*, N. Guarino, Editor. 1998, IOS Press: pp. 3-15.

22. Smith, B., *Mereotopology: A Theory of Parts and Boundaries*. Data and Knowledge Engineering, 1996. 20: pp. 287-303.

23. Gibson, J., *The Theory of Affordances*, in *Perceiving, Acting, and Knowing - Toward an Ecological Psychology*, R. Shaw and J. Bransford, Editors. 1977, Lawrence Erlbaum Associates: pp. 67-82.

24. Rugg, R., M. Egenhofer, and W. Kuhn, *Formalizing Behavior of Geographic Feature Types.* Geographical Systems, 1997. 4(2): pp. 159-180.
25. Goguen, J.A., *An Introduction to Algebraic Semiotics, with Application to User Interface Design,* in *Computation for Metaphors, Analogy and Agents,* C. Nehaniv, Editor. 1999, Springer, Lecture Notes in Artificial Intelligence 1562: pp. 242–291.
26. Harnad, S., *The Symbol Grounding Problem.* Physica D, 1990. 42: pp. 335-346.
27. Zlatev, J., *Situated embodiment: Studies in the emergence of spatial meaning.* 1997, Stockholm: Gotab.
28. Goguen, J.A., *Ontology, Society, and Ontotheology,* in *Formal Ontology in Information Systems, Proceedings of the Third International Conference (FOIS 2004),* A. Varzi and L. Vieu, Editors. 2004, IOS Press, 114: pp. 95-103.
29. Fensel, D. and C. Bussler, *The Web Service Modeling Framework WSMF.* Electronic Commerce: Research and Applications, 2002(1): pp. 113-137.
30. Parnas, D.L., *A Technique for Software Module Specification with Examples.* Communications of the ACM, 1972. 15(5): pp. 1053-1058.
31. Cook, S. and J.D. Daniels, *Designing Object Systems: Designing Object Oriented Modelling with Syntropy.* 1994: Prentice Hall.
32. Buehler, K., ed. *OpenGIS Reference Model.* 2003, Open Geospatial Consortium (OGC).
33. Kottmann, C., *Semantics and Information Communities,* in *The OpenGIS™ Abstract Specification,* C. Kottmann, Editor. Open GIS Consortium (OGC), 1999. http://www.opengeospatial.org/docs/99-114.pdf.
34. Riedemann, C. and C. Timm, *Services for Data Integration.* Data Science Journal, 2003. 2(26): pp. 90-99.
35. ISO, *ISO 19107 - Spatial Schema.* ISO TC 211, 2002.
36. Barr, M. and C. Wells, *Category Theory for Computing Science.* 1990: Prentice Hall.
37. Ehrig, H. and B. Mahr, *Fundamentals of Algebraic Specification.* 1985: Springer.
38. Woodcock, J. and M. Loomes, *Software Engineering Mathematics.* The SEI Series in Software Engineering. 1989: Addison Wesley.
39. Arpinar, I.B., et al., *Geospatial Ontology Development and Semantic Analytics,* ed. J.P. Wilson and A.S. Fotheringham. Handbook of Geographic Information Science. 2005 (in print): Blackwell Publishing.
40. Schade, S., et al. *Comparing Approaches for Semantic Service Description and Matchmaking.* in *3rd Int. Conf. on Ontologies, Data Bases, and Applications of Semantics for Large Scale Information Systems (ODBASE 2004).* 2004. Larnaca, Cyprus: Springer, Lecture Notes in Computer Science 3291: pp. 1062-1079.
41. Probst, F., et al. *Connecting ISO and OGC Models to the Semantic Web (Extended Abstract).* in *Third International Conference on Geographic Information Science.* 2004. Adelphi, MD, USA: pp. 181-184.
42. 42. Kavouras, M. and M. Kokla, *A Method for the Formalization and Integration of Geographic Categorizations.* International Journal of Geographical Information Science, 2002. 16(5): pp. 439-453.
43. Frank, A., *Ontology for spatio-temporal Databases,* in *Spatiotemporal Databases: The Chorochronos Approach,* M.e.a. Koubarakis, Editor. 2003, Springer, 2520: pp. 9-77.
44. Lehnert, K., Su, Y., Langmuir, C.H., Sarbas, B., Nohl, U., *A global geochemical database structure for rocks.* Geochemistry, Geophysics, Geosystems, 2000. 1(May).
45. Worboys, M. and K. Hornsby, *From Objects to Events: GEM, the Geospatial Event Mode,* in *Geographic Information Science. Third International Conference, GIScience 2004,* M.J. Egenhofer, C. Freksa, and H.J. Miller, Editors. 2004, Springer, Lecture Notes in Computer Science 3234: pp. 327-343.
46. Bennett, B. and M. Cristani, eds. *Spatial Cognition and Computation: special issue on spatial vagueness, uncertainty and granularity.* Spatial Cognition and Computation, ed. T. Cohn and S. Hirtle. Vol. 3(2-3). 2004, Springer.

47. Stell, J.G., *Granularity in Change over Time*, in *Foundations of Geographic Information Science*, M. Duckham, M. Goodchild, and M. Worboys, Editors. 2003, Taylor & Francis: pp. 95-115.
48. Fonseca, F., et al., *Semantic Granularity in Ontology-Driven Geographic Information Systems*. Annals of Mathematics and Artificial Intelligence, 2002. 36(1-2): pp. 121-151.
49. Bishr, Y., *Overcoming the Semantic and Other Barriers to GIS Interoperability*. IJGIS, 1998. 12(4): pp. 299-314.
50. Sycara, K., et al., *Dynamic Service Matchmaking Among Agents in Open Information Environments*. SIGMOD Record, 1999. 28(1): pp. 47-53.
51. Klien, E., et al. *An Architecture for Ontology-Based Discovery and Retrieval of Geographic Information*. in *7th Conference on Geographic Information Science (AGILE 2004)*. 2004. Heraklion, Greece: Crete University Press: pp. 179-188.
52. Egenhofer, M. *Toward the Semantic Geospatial Web*. in *10th ACM International Symposium on Advances in Geographic Information Systems (ACM-GIS)*. 2002. McLean, VA: pp. 1-4.
53. Lutz, M., *Operation Ontologies for Semantic Discovery and Composition of Geoprocessing Services*. Münster Semantic Interoperability Lab (MUSIL), 2005.
54. Riedemann, C., *Naming Topological Operators at GIS User Interfaces*. Münster Semantic Interoperability Lab (MUSIL), 2005.
55. Bernstein, A. and M. Klein. *Towards High-Precision Service Retrieval*. in *The Semantic Web - First International Semantic Web Conference (ISWC 2002)*. 2002. Sardinia, Italy: pp. 84-101.
56. Iliffe, J.C., *Datums and Map Projections*. 2000.
57. Vinoski, S., *Web Services Interaction Models, Part 1: Current Practice*. IEEE Internet Computing, 2002(May-June 2002): pp. 89-91.
58. Bernard, L., et al. *Interoperability in GI Service Chains - The Way Forward*. in *6th AGILE Conference on Geographic Information Science*. 2003. Lyon, France.
59. Probst, F. and M. Lutz. *Giving Meaning to GI Web Service Descriptions*. in *2nd International Workshop on Web Services: Modeling, Architecture and Infrastructure (WSMAI-2004)*. 2004. Porto, Portugal.
60. Rodríguez, A. and M. Egenhofer, *Comparing geospatial entity classes: an asymmetric and context-dependent similarity measure*. International Journal of Geographical Information Science, 2004. 18(3): pp. 229-256.
61. Rodríguez, A. and M. Egenhofer. *Putting Similarity Assessment into Context: Matching Functions with the User's Intended Operations*. in *Modeling and Using Context, CONTEXT'99*. 1999. Trento, Italy: Springer-Verlag, Lecture Notes in Computer Science 1688: pp. 310-323.
62. Gärdenfors, P., *How to Make the Semantic Web More Semantic*, in *Formal Ontology in Information Systems, Proceedings of the Third International Conference (FOIS 2004)*, A. Varzi and L. Vieu, Editors. 2004, IOS Press, 114: pp. 17-34.
63. Fabrikant, S.I. and B.P. Buttenfield, *Formalizing Semantic Spaces for Information Access*. Annals of the Association of American Geographers, 2001. 91: pp. 263-280.
64. Kuhn, W., *Semantic Reference Systems*. International Journal of Geographic Information Science (Guest Editorial), 2003. 17(5): pp. 405-409.
65. Chrisman, N., *Exploring Geographic Information Systems*. 2nd ed. 2002: Wiley.
66. Goodchild, M., *Geographical data modeling*. Computers and Geosciences, 1992. 18(4): pp. 401-408.
67. Kuhn, W. and M. Raubal. *Implementing Semantic Reference Systems*. in *AGILE 2003 - 6th AGILE Conference on Geographic Information Science*. 2003. Lyon, France: Presses Polytechniques et Universitaires Romandes: pp. 63-72.
68. Berners-Lee, T., J. Hendler, and O. Lassila, *The Semantic Web*, in *Scientific American*, 2001. pp. 34-43.

69. Raubal, M., *Formalizing Conceptual Spaces*, in *Formal Ontology in Information Systems, Proceedings of the Third International Conference (FOIS 2004)*, A. Varzi and L. Vieu, Editors. 2004, IOS Press, 114: pp. 153-164.
70. Doerr, M., J. Hunter, and C. Lagoze, *Towards a Core Ontology for Information Integration*. Journal of Digital information, 2003. 4(1).
71. Frank, A., *Pragmatic Information Content: How to Measure the Information in a Route Description*, in *Foundations of Geographic Information Science*, M. Duckham, M. Goodchild, and M. Worboys, Editors. 2003, Taylor & Francis: pp. 47-68.
72. Oakley, T., *Image Schema*, in *Handbook of Cognitive Linguistics*, D. Geeraerts and H. Cuyckens, Editors. (in press), Oxford University Press. http://www.cwru.edu/artsci/engl/oakley/image_schema.pdf.
73. Johnson, M., *The Body in the Mind: The Bodily Basis of Meaning, Imagination, and Reason*. 1987: The University of Chicago Press.
74. Lakoff, G., *The Invariance Hypothesis: is abstract reason based on image-schemas?* Cognitive Linguistics, 1990. 1(1): pp. 39-74.
75. Stevens, S.S., *On the Theory of Measurement*. Science, 1946. 103(2684): pp. 677-680.
76. Klein, F., *Vergleichende Betrachtungen über neuere geometrische Forschungen*. 1872: Verlag Andreas Deichert.

Spherical Topological Relations

Max J. Egenhofer

National Center for Geographic Information and Analysis,
Department of Spatial Information Science and Engineering,
Department of Computer Science,
University of Maine, Orono, ME 044690-5711, USA
http://www.spatial.maine.edu/~max
max@spatial.maine.edu

Abstract. Analysis of global geographic phenomena requires non-planar models. In the past, models for topological relations have focused either on a two-dimensional or a three-dimensional space. When applied to the surface of a sphere, however, neither of the two models suffices. For the two-dimensional planar case, the eight binary topological relations between spatial regions are well known from the 9-intersection model. This paper systematically develops the binary topological relations that can be realized on the surface of a sphere. Between two regions on the sphere there are three binary relations that cannot be realized in the plane. These relations complete the conceptual neighborhood graph of the eight planar topological relations in a regular fashion, providing evidence for a regularity of the underlying mathematical model. The analysis of the algebraic compositions of spherical topological relations indicates that spherical topological reasoning often provides fewer ambiguities than planar topological reasoning. Finally, a comparison with the relations that can be realized for one-dimensional, ordered cycles draws parallels to the spherical topological relations.

1 Introduction

GIS applications that deal with phenomena that spread across the entire globe need semantic models of spatial relations that consider the particular properties of the sphere (Usery 2002). For example, an atmospheric scientist studying global warming needs a spherical geometric representation of the Earth to model accurately the dynamic processes of long-term climate change. Likewise spatio-temporal analyses of the worldwide diffusion of diseases benefit from models based on the sphere. The sphere is a two-dimensional space that is embedded in a three-dimensional setting such that it separates the embedding universe (typically $\mathbb{R}^3$) into two disconnected parts—the interior and the exterior of a globe. Models for qualitative spatial relations, particularly topological relations, have received much attention in the GIS and spatial-database literature over the last decade (Egenhofer and Franzosa 1991; Hadzilacos and Tryfona 1992; Smith and Park 1992; Clementini *et al.* 1993; Cui *et al.* 1993; Clementini *et al.* 1994; Egenhofer *et al.* 1994; Clementini *et al.* 1995; Egenhofer and Franzosa 1995; Papadias *et al.* 1995; Winter 1995; Cohn and Gotts 1996; Papadimitriou *et al.* 1996; Clementini and di Felice 1997; Billen *et al.* 2002). Implementations in commercial GISs (e.g., Intergraph's MGA and ESRI's SDE) and spatial database

S. Spaccapietra and E. Zimányi (Eds.): Journal on Data Semantics III, LNCS 3534, pp. 25 – 49, 2005.
© Springer-Verlag Berlin Heidelberg 2005

systems (e.g., Oracle10g Spatial) exist and several standards and drafts of standards have incorporated various versions (e.g., SAIF, ISO TC/211, OGC's Simple Feature Specification). Most of the focus has been on relations in two-dimensional, occasionally three-dimensional space (Pigot 1991; Hazelton *et al.* 1992), but little attention has been paid to investigating models of such qualitative spatial relations on the surface of a sphere.

This paper derives the set of binary topological relations that can be found between two regions on the sphere $\mathbb{P}^2$, with $\mathbb{P} \subset \mathbb{R}$ such that $\mathbb{P}$ is connected and $\min(\mathbb{P}) = \max(\mathbb{P})$. For this purpose, this paper employs the 9-intersection (Egenhofer and Herring 1991) as a model for binary topological relations. It further analyzes the qualitative reasoning power of this set of relations in terms of its conceptual neighborhoods—a measure for the similarity of relations—and its compositions—a foundation for symbolic reasoning in terms of a relation algebra. Two comparisons are made throughout the paper. The first benchmark is the set of topological relations that can be realized in the two-dimensional plane $\mathbb{R}^2$. The second benchmark is the transition from a one-dimensional space $\mathbb{R}^1$, as used for temporal reasoning, to a cyclic one-dimensional space $\mathbb{P}^1$. Within these settings, we are particularly interested in answering the following four questions:

- Does the mapping from $\mathbb{R}^2$ onto $\mathbb{P}^2$ reduce the number of relations found in $\mathbb{R}^2$ but not in $\mathbb{P}^2$?
- Do additional binary topological relations exist in $\mathbb{P}^2$ that cannot be realized in $\mathbb{R}^2$?
- Are the conceptual neighborhoods of all relations in $\mathbb{P}^2$ a consistent theoretical framework for organizing binary spherical topological relations according to their similarity?
- Are inferences based upon the composition of topological relations in $\mathbb{P}^2$ less crisp than in $\mathbb{R}^2$?

The significance of the findings from this investigation is twofold. First, it is of immediate interest for a spatial inference engine to know what types of global spatial relations may be realized on a sphere but cannot be found in a plane. Such knowledge will provide the basis for future spatial query processors that apply to three-dimensional spatial data models or augment early versions, such as the Geodetic DataBlade (IBM 2002), which offers a three-dimensional data model that features only three binary topological relations—*inside*, *intersect*, and *outside*. Second, finding parallels between relations in one-dimensional and two-dimensional spaces—as well as parallels in the transition from linear to cyclic spaces—may give us new insights about the scalability of certain spatial properties. The latter is part of investigations into spatial theories and forms a fundamental aspect of any such formalization in geographic information science.

The remainder of this paper is structured as follows: Section 2 compares similarities and differences between a cyclic one-dimensional and spherical two-dimensional space, followed in Section 3 by a summary of the model for binary topological relations in $\mathbb{R}^2$. Section 4 develops the set of binary topological relations that can be realized on a sphere and compares the results with the relations realized in a cyclic one-dimensional space. Section 5 proves the completeness of this set of spherical topological relations. Section 6 derives systematically the compositions of spherical topological relations and compares the inference power with that of the topological relations in $\mathbb{R}^2$. Conclusions in Section 7 summarize the major findings.

2   Similarities Between Cyclic One-Dimensional and Spherical Two-Dimensional Relations

Until recently the embedding space for one-dimensional (temporal) relations has been primarily the linear timeline that corresponds to the real numbers $\mathbb{R}^1$, while the setting that gives rise to cyclic temporal relations (i.e., relations that are embedded in a cyclic, one-dimensional space, denoted by $\mathbb{P}^1$) has been largely ignored. Cyclic one-dimensional relations expose the following properties (Hornsby *et al.* 1999; Balbiani and Osmani 2000):

- one pair of relations that can be realized in $\mathbb{R}^1$ collapses to a single relation in $\mathbb{P}^1$;
- in $\mathbb{P}^1$ additional binary relations exist that cannot be realized in $\mathbb{R}^1$; and
- the conceptual neighborhoods of the relations in $\mathbb{P}^1$ form a framework for a systematic analysis of the completeness of the relations.

We want to verify that similar conclusions can be drawn when the embedding two-dimensional space $\mathbb{R}^2$ gets warped into the surface of a sphere $\mathbb{P}^2$, much like forming a one-dimensional cycle $\mathbb{P}^1$ from a linear, one-dimensional space $\mathbb{R}^1$. Investigations of these comparisons are enabled by the existence of two similar frameworks for organizing such spatial relations in $\mathbb{R}^1$ and $\mathbb{R}^2$. In both cases, the basic sets of relations in $\mathbb{R}^1$ (Allen 1983) and $\mathbb{R}^2$ (Egenhofer and Franzosa 1991) have been widely popular and provide foundations for studies of relations in $\mathbb{P}^1$ and $\mathbb{P}^2$, respectively. The analogy between cyclic one-dimensional relations and spherical two-dimensional relations stems from common properties found in both embedding spaces.

One property is that both types of relations are located in a space that is embedded in a higher-dimensional space—at least a two-dimensional plane for cycles and at least a three-dimensional space for spheres. We call such an embedding space the *co-space*. If the *co-dimension*—the difference between the dimension of the co-space and the dimension of the reference object's space—is equal to 1, then the reference space acts as a Jordan curve (or its higher-dimensional equivalents), separating the co-space into two disconnected parts, an inner and an outer co-space. This property holds for the cyclic one-dimensional space as well as for the spherical two-dimensional space. Cyclic one-dimensional relations and spherical two-dimensional relations both attempt to capture qualitative information (Hernández 1994). Such information typically relies on properties that are invariant under certain types of transformations.

Despite these commonalities, there are some significant differences between a one-dimensional and a two-dimensional embedding, which make it impossible to generalize all findings from the one-dimensional space and apply them to a two-dimensional space. At the outset, the two approaches differ in the way they make use of the order of the space. Whereas the set of one-dimensional relations that disregards the order (Pullar and Egenhofer 1988) typically finds its applications in higher-dimensional spaces (e.g., cartographic applications featuring line relations with co-dimension 1),

Allen's interval relations are tailored to representations of time and, therefore, exploit the order of $\mathbb{R}^1$, which is based on an order relation ($\leq$) with the usual algebraic properties of reflexivity, antisymmetry, and transitivity. With the transition from a linearly ordered one-dimensional space to a cyclically ordered one-dimensional space, the orientation is reduced to a less powerful relation that lacks transitivity. This difference in properties has implications on what relations can be distinguished. While A *before* B and A *after* B are two distinct relations in $\mathbb{R}^1$, they blend in $\mathbb{P}^1$ into a single relation, *disjoint* (Hornsby *et al.* 1999). On the other hand, the difference between A *meets* B and A *metBy* B, which is also due to the underlying order relation, is retained in $\mathbb{P}^1$ due to the orientation of the cycle. In $\mathbb{R}^2$ and $\mathbb{P}^2$, however, the setting is different. The orientation of a two-dimensional space has no observable influence on the choice of topological relations—although an enhancement of topological relations with cardinal directions provides an extension that offers additional expressive power (Sharma 1999). Therefore, one could expect that the transition from $\mathbb{R}^2$ and $\mathbb{P}^2$ does not offer the same contraction in a pair of relations as the transition from $\mathbb{R}^1$ to $\mathbb{P}^1$ does.

Another important difference relates to a property of the boundaries of a one-dimensional and a two-dimensional object. In a one-dimensional space the basic object of interest is an interval, which is a non-empty, closed, connected, and proper subset of a one-dimensional space. The boundary of such an interval forms a separation, that is, in order to connect all parts of the boundary it is necessary to traverse the interval's interior or exterior. On the other hand, in a two-dimensional space $\mathbb{R}^2$ the basic object is a *region*—defined as a non-empty proper subset of a connected topological space such that the region's interior is connected and the region is identical to the closure of the region's interior (Egenhofer and Franzosa 1992). It is closed, bounded, homogeneously two-dimensional, and homeomorphic to a 2-disk. Unlike the interval's boundary, a region's boundary is connected, that is, any parts of its boundary can be connected by a line without a need to traverse the region's interior or exterior. This difference between one-dimensional and two-dimensional elements in their corresponding spaces already led to different properties of one pair of topological relations. In 1-D the *overlap* relation has an empty boundary-boundary intersection, while in 2-D the corresponding relation requires the two boundaries to intersect (Egenhofer *et al.* 1993).

These differences indicate that the transition from $\mathbb{R}^1$ to $\mathbb{P}^1$ is not fully parallel to the transition from $\mathbb{R}^2$ to $\mathbb{P}^2$. Still a significant similarity exists between the two scenarios, and we study them subsequently.

3   Topological Relations in $\mathbb{R}^2$

The 9-intersection defines binary topological relations between two regions, A and B, in terms of A's interior (A°), boundary (∂A), and exterior (A^-) with B's interior (B°), boundary (∂B), and exterior (B^-) (Egenhofer and Herring 1991). The nine intersections between these six object parts describe a topological relation and can be concisely represented by a 3×3-matrix, called the *9-intersection* (Equation 1).

$$I_9 = \begin{pmatrix} A^\circ \cap B^\circ & A^\circ \cap \partial B & A^\circ \cap B^- \\ \partial A \cap B^\circ & \partial A \cap \partial B & \partial A \cap B^- \\ A^- \cap B^\circ & A^- \cap \partial B & A^- \cap B^- \end{pmatrix} \tag{1}$$

Topological invariants of these nine intersections (i.e., properties that are preserved under topological transformations) are used to categorize topological relations. Examples of topological invariants, applicable to the 9-intersection, are the content (i.e., emptiness or non-emptiness) of a set, the dimension, and the number of separations (Egenhofer and Franzosa 1995). The content invariant is the most general criterion, because other invariants can be considered refinements of non-empty intersections. By considering the values empty ($\emptyset$) and non-empty ($\neg\emptyset$) for each of the nine intersections, one can distinguish $2^9 = 512$ binary topological relations. Eight of these 512 relations can be realized between two regions embedded in $\mathbb{R}^2$. They are subsequently referred to as the $\mathbb{R}^2$-relations. Although the subset of the four intersections of the regions' interiors and boundaries—called the 4-intersection—is sufficient to distinguish the eight $\mathbb{R}^2$-relations, the 9-intersection captures critical information for making inferences about combinations of topological relations (Egenhofer 1994a).

4   Topological Relations on a Sphere

We develop the spherical topological relations in two steps. First, we build on the eight $\mathbb{R}^2$-relations and examine whether they can be realized in $\mathbb{P}^2$ (Section 4.1), before we investigate what relations are particular to $\mathbb{P}^2$ and, therefore, beyond the set of eight $\mathbb{R}^2$-relations (Section 4.2).

The definition of a region in $\mathbb{P}^2$ is identical to that of a region used for the study of topological relations in the plane (Egenhofer and Franzosa 1992) and, therefore, allows direct comparisons. A region has a non-empty interior, a non-empty boundary, and a non-empty exterior, and interior and exterior are simply connected. This definition eliminates some borderline cases of objects that may occur on spheres but are not subject of the present study, such as the entire sphere (because the boundary and the exterior would be empty), a sphere with a crack (because the exterior would be empty), and subsets of $\mathbb{P}^2$ with disconnected exteriors (e.g., regions with holes) and disconnected interiors (e.g., regions with separations).

While the union of two regions in the plane cannot cover the entire embedding space $\mathbb{R}^2$ (Egenhofer and Franzosa 1992), it is possible on the sphere that the union of two regions is identical to $\mathbb{P}^2$. A study of the properties of regions on the sphere (Gotts 1996)—not relations between regions—used the region-connected calculus (Randell *et al.* 1992), a formalism that yields results comparable to those of the 9-intersection.

4.1   Realizability of $\mathbb{R}^2$-Relations in $\mathbb{P}^2$

The first question addresses whether all of the eight $\mathbb{R}^2$-relations can be found in $\mathbb{P}^2$ and if so, whether they can be distinguished uniquely in $\mathbb{P}^2$ as well. A straightforward task is to warp a two-dimensional plane, with two regions on it, so that it forms a sphere. On that sphere we find that all eight region-region relations from $\mathbb{R}^2$ have 1:1 corresponding topological relations (Figure 1). Since the same underlying assumptions of the 9-intersection apply to $\mathbb{R}^2$ and $\mathbb{P}^2$—a connected boundary separates a simply connected exterior from a simply connected interior—the 9-intersection serves as a valid model to distinguish these eight relations in $\mathbb{P}^2$ as well. Therefore, we have found the answer to the initial question about the scalability of cyclic relations:

- While Allen's temporal interval relations, which rely on an order relation, do not scale up immediately from $\mathbb{R}^1$ to $\mathbb{P}^1$—in this process one pair of $\mathbb{R}^1$-relations gets merged into a single $\mathbb{P}^1$-relation (Hornsby *et al.* 1999)—the transition from $\mathbb{R}^2$ to $\mathbb{P}^2$ does not have a similar impact on the topological relations, as it retains all $\mathbb{R}^2$-relations in $\mathbb{P}^2$.

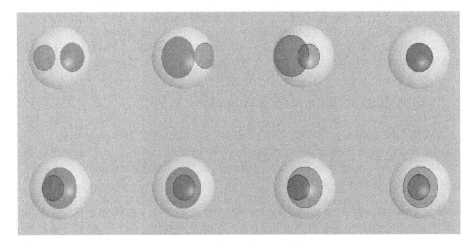

Fig. 1. Examples of the eight topological relations that can be realized in $\mathbb{R}^2$ and in $\mathbb{P}^2$

4.2   Exclusively Spherical Relations

What binary topological relations does a sphere reveal that $\mathbb{R}^2$ would not permit? To answer this question, we start with the topological relation that occurs when two half-spheres are attached to each other so that their union forms a complete surface of a sphere (Figure 2a). In this case, the two boundaries coincide, while each object's interior coincides with the other object's exterior. The same relation holds for any configuration homeomoprhic to this setting with two half-spheres. We call this relation *attach*. To distinguish *attach* from *meet*, we need to use the 9-intersection, because the difference between the two relations is captured by the way the boundaries lay with respect to the exteriors, which is a property that cannot be captured by the 4-intersection (Equation 2). The *attach* relation cannot be realized between two regions in $\mathbb{R}^2$, because for two regions in the plane the coincidence of the two boundaries would imply a coincidence of the two interiors, which represents the relation *equal*.

$$\begin{pmatrix} \varnothing & \varnothing & \neg\varnothing \\ \varnothing & \neg\varnothing & \varnothing \\ \neg\varnothing & \varnothing & \varnothing \end{pmatrix} \tag{2}$$

Another spherical topological relation occurs if the *attach* relation is deformed such that parts, but not all, of the boundary of each region runs through the interior of the other region (Figure 2b). This relation is called *entwined*. Again the 9-intersection is needed to describe this relation, because the 4-intersection alone cannot distinguish it from *overlaps* (Equation 3). *Entwined* cannot be realized between two regions in $\mathbb{R}^2$, because for two regions in the plane the inclusion of one region's boundary in the other region's closure (such that it intersects with the interior and boundary) would imply the relation *covers*.

$$\begin{pmatrix} \neg\varnothing & \neg\varnothing & \neg\varnothing \\ \neg\varnothing & \neg\varnothing & \varnothing \\ \neg\varnothing & \varnothing & \varnothing \end{pmatrix} \tag{3}$$

The third exclusively spherical relation is one in which each region's boundary is located completely in the interior of the other region's interior, while each region's exterior is located completely in the other region's interior (Figure 2c). This relation is called *embrace*. It is the only $\mathbb{P}^2$-relation that can be distinguished with the 4-intersection from the eight $\mathbb{R}^2$-relations (Equation 4). It is impossible, however, to realize it between two regions in $\mathbb{R}^2$, because for two regions in the plane the complete inclusion of one region's boundary in the other region's interior implies the relation *contains*.

$$\begin{pmatrix} \neg\varnothing & \neg\varnothing & \neg\varnothing \\ \neg\varnothing & \varnothing & \varnothing \\ \neg\varnothing & \varnothing & \varnothing \end{pmatrix} \tag{4}$$

All exclusively spherical relations are such that the union of the two regions forms the entire sphere. This property does not hold for any of the eight topological relations that were projected from $\mathbb{R}^2$ into $\mathbb{P}^2$, nor did it hold for any of the eight region-region relations in $\mathbb{R}^2$. Furthermore, all exclusively spherical relations are symmetric, because their 9-intersection matrices are symmetric with respect to the main diagonal.

With the identification of these three exclusively spherical relations, we have found the answer to the second question about the scalability of cyclic relations:

- Similar to the mapping from $\mathbb{R}^1$ to $\mathbb{P}^1$, the mapping from $\mathbb{R}^2$ to $\mathbb{P}^2$ gives rise to new binary topological relations between two regions that cannot be found between two regions in $\mathbb{R}^2$.

It is an interesting observation that the three exclusively spherical relations are such that the union of the two regions coincides with $\mathbb{P}^2$. One might argue that these new relations could have been obtained in $\mathbb{R}^2$ as well if one allowed two regions to

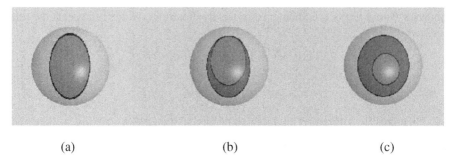

(a)   (b)   (c)

Fig. 2. Examples of the three topological relations that can be realized only on the sphere $\mathbb{P}^2$: (a) *attach*, (b) *entwined*, and (c) *embrace*

be such that their union forms $\mathbb{R}^2$. Such an approach, however, would require a modification of the basic definition of a spatial region (Egenhofer and Franzoa 1992) to include non disk-like configurations such as a half plane. In order to stay within the scope of the established setting for geographic applications, such modifications are not desired.

4.3   Completeness of the Set of Topological Relations in $\mathbb{P}^2$

Analogous to the discovery of the topological relations in $\mathbb{R}^2$ (Egenhofer and Herring 1991), we prove the completeness of this set of spherical topological relations by examining what 9-intersection combinations cannot be realized between two regions on the sphere. We capture impossible relations as constraints among the elements of the 9-intersection matrix. Some of these constraints are common to regions in $\mathbb{R}^2$, while others that applied to $\mathbb{R}^2$ do not hold true in $\mathbb{P}^2$.

Ten constraints apply to the interactions between interiors, boundaries, and exteriors for two regions on the sphere (Equations 5a-j). They also apply in the reverse direction, from B to A, by exchanging systematically A and B in Equations 5a-k.

Constraint 1:   The two interiors A° and B° cannot be disjoint at the same time as A° is disjoint from the exterior of B (Equation 5a).

$$\nexists\ A,B:\quad A^\circ \cap B^\circ = \varnothing\ \wedge\ A^\circ \cap B^- = \varnothing \tag{5a}$$

Proof: A° and B° must be non-empty (Section 4). Since at least one part of B must be non-empty, it follows that if $A^\circ \cap B^\circ$ is empty, A° must have a non-empty intersection with ∂B or B^-. Assume that $A^\circ \cap B^-$ is empty, then A° would have to be totally included in ∂B, which is impossible. On the other hand, if $A^\circ \cap \partial B$ is empty, then A° would have to be totally included in B^-, that is, $A^\circ \cap B^- = \neg\varnothing$, which contradicts $A^\circ \cap B^- = \varnothing$.   ∴

Constraint 2:   The two interiors A° and B° cannot be disjoint at the same time as A° intersects with B's boundary (Equation 5b)

$$\nexists\ A,B:\quad A^\circ \cap B^\circ = \varnothing\ \wedge\ A^\circ \cap \partial B = \neg\varnothing \tag{5b}$$

Proof: Detailed proof was included in Egenhofer and Franzosa (1991).   ∴

Constraint 3:   A's interior $A°$ cannot intersect with B's boundary at the same time as $A°$ is disjoint from B's exterior (Equation 5c).

$$\nexists\ A,B:\ \ A°\cap\partial B=\neg\varnothing\ \wedge\ A°\cap B^-=\varnothing \tag{5c}$$

Proof: Follows from proof of Constraint 2.   ∴

Constraint 4: A's interior cannot be disjoint from B's exterior B^- at the same time as A's boundary intersects with B^- (Equation 5d).

$$\nexists\ A,B:\ \ A°\cap B^-=\varnothing\ \wedge\ \partial A\cap B^-=\neg\varnothing \tag{5d}$$

Proof: Follows from proof of Constraint 2.   ∴

Constraint 5: A's interior $A°$ cannot intersect with B's interior at the same time as $A°$ is disjoint from B's boundary and $A°$ intersects with B's exterior (Equation 5e).

$$\nexists\ A,B:\ \ A°\cap B°=\neg\varnothing\ \wedge\ A°\cap\partial B=\varnothing\ \wedge\ A°\cap B^-=\neg\varnothing \tag{5e}$$

Proof: The three parts of B—$B°$, ∂B, and B^-—form a complete partition of space. They are also arranged such that $B°$ is adjacent to ∂B and ∂B is adjacent to B^-. Since ∂B forms a Jordan curve, separating $B°$ from B^-, there is no connection from $B°$ to B^- without going through ∂B. Therefore, if $A°$ has non-empty intersections with $B°$ and with B^-, it must have a non-empty intersection with ∂B as well, which contradicts $A°\cap\partial B=\varnothing$.   ∴

Constraint 6: A's boundary cannot intersect with B's exterior B^- at the same time as A's exterior is disjoint from B^- (Equation 5f).

$$\nexists\ A,B:\ \ A^-\cap B^-=\varnothing\ \wedge\ \partial A\cap B^-=\neg\varnothing \tag{5f}$$

Proof: Analog to proof of constraint 2, replacing A^- and B^- with $A°$ and $B°$, respectively.   ∴

Constraint 7: A's boundary ∂A cannot be disjoint from B's interior at the same time as ∂A is disjoint from B's boundary and ∂A is disjoint from B's exterior (Equation 5g).

$$\nexists\ A,B:\ \ \partial A\cap B°=\varnothing\ \wedge\ \partial A\cap\partial B=\varnothing\ \wedge\ \partial A\cap B^-=\varnothing \tag{5g}$$

Proof: The three parts of B—$B°$, ∂B, and B^-—form a complete partition of space. Also, ∂A must be non-empty. Therefore, ∂A must have a non-empty intersection with at least one part of B.   ∴

Constraint 8: A's boundary ∂A cannot intersect with B's interior at the same time as ∂A is disjoint from B's boundary and ∂A intersects with B's exterior (Equation 5h).

$$\nexists\ A,B:\ \ \partial A\cap B°=\neg\varnothing\ \wedge\ \partial A\cap\partial B=\varnothing\ \wedge\ \partial A\cap B^-=\neg\varnothing \tag{5h}$$

Proof: Follows from proof of constraint 5. $\therefore$

Constraint 9:   A 's exterior A^- cannot intersect with B 's interior at the same time as A^- is disjoint from B 's boundary and A^- is disjoint from B 's exterior (Equation 5i).

$$\not\exists\ A,B:\quad A^-\cap B^\circ = \neg\varnothing\ \wedge\ A^-\cap\partial B=\varnothing\ \wedge\ A^-\cap B^- = \neg\varnothing \tag{5i}$$

Proof: Follows from proof of constraint 5. $\therefore$

Constraint 10:   A 's exterior A^- cannot be disjoint from B 's interior at the same time as A^- is disjoint from B 's boundary and A^- is disjoint from B 's exterior (Equation 5j).

$$\not\exists\ A,B:\quad A^-\cap B^\circ = \varnothing\ \wedge\ A^-\cap\partial B=\varnothing\ \wedge\ A^-\cap B^- = \varnothing \tag{5j}$$

Proof: Follows from proof of constraint 7. $\therefore$

With the help of a Prolog program we determined the set of 9-intersections that do not violate any of these constraints. The resulting set consists of the 9-intersections of the eleven spherical topological relations determined in Sections 4.2 and 4.3. The ten constraints are not redundant, as could be demonstrated by an attempt to remove each constraint from the set of ten and recalculate the set of possible relations. For all possible combinations of selecting only nine out of ten constraints, the resulting set of 9-intersection combinations was larger than the set obtained by using all ten constraints. A different—possibly smaller, but equivalent—set of constraints might be found in the future, but it would not change the purpose or the confirmation of this set's completeness.

5   Similarity Among Topological Relations in $\mathbb{P}^2$

Conceptual neighborhoods have been used successfully in the analysis of sets of relations for similarity (Egenhofer and Al-Taha 1992; Freksa 1992; Egenhofer and Mark 1995). The conceptual neighborhood graph captures for each relation those relations that are conceptually closest to it. Two relations are neighbors if a continuous transformation can be performed between the two relations without the need to go through a third relation. Since relations to be related typically lack a total order, their conceptual neighborhoods are used as the primary tool to provide insights about the closeness or similarity of the relations (Bruns and Egenhofer 1996). They also provide a foundation for the selection of appropriate natural-language terminology when people communicate with information systems (Mark and Egenhofer 1994).

The conceptual neighborhood for the eight topological relations in $\mathbb{R}^2$, denoted by N_8, forms a connected graph in which pairs of relations that are connected directly by an edge correspond to transitions that can be obtained by applying topological transformations—translation, rotation, or scaling—to one or both objects. On the other hand, pairs of relations that are not directly connected cannot be obtained through such topological transformations. Further connections—from *inside* to *equal* and from *equal* to *contains*—could be established by considering a scaling that changes boundaries at all points simultaneously. Likewise, for objects with the same size, shape, and orientation a direct transition from *overlap* to *equal* could be established. Such additional links, however, would not change the overall layout and properties of the conceptual neighborhood graph. N_8 has a vertical symmetry axis that coincides with all symmetric relations and the mirror images along this axis correspond to pairs of converse relations.

The conceptual neighborhood for the eleven topological relations in $\mathbb{P}^2$, denoted by N_{11}, can be derived with the same rationale as N_8 (Egenhofer and Al-Taha 1992). For each pair of spherical topological relations, r_a and r_b, the number of differences in the 9-intersection at corresponding intersections, denoted by $I_{r_a}[i,j]$ and $I_{r_b}[i,j]$, provides a metric for the topological difference of the relations (Equation 6), where the difference is 0 between two empty elements, 0 between two non-empty elements, and 1 between an empty and a non-empty element, as well as between a non-empty and an empty element. Therefore, the sum over all nine interior-, boundary-, and exterior-intersections, denoted by $\tau(r_a, r_b)$, is a cumulative, equal-weight difference value.

$$\tau(r_a, r_b) = \sum_{i=\circ}^{-} \sum_{j=\circ}^{-} (I_{r_a}[i,j] - I_{r_b}[i,j]) \qquad (6)$$

The conceptual neighbors of a relation r_a comprise the set of those relations r_x with the smallest non-zero difference $\tau(r_a, r_x)$ (Table 1). This constraint is not necessarily symmetric, because a relation r_b may be found to be among the least different relations from r_a without the requirement that r_a is among the least different relations from r_b. Since the conceptual neighborhood graph is a non-directed graph, these differences are not captured in N_{11}.

The conceptual neighborhood graph for the eleven topological relations that can be realized on the sphere shows how the three spherical relations fan off from the relations *meet* and *overlap* in the upper half of the graph (Figure 3a). There is no connection to any of the three spherical relations in the lower half of the graph. The six relations located in the upper half of the graph, denoted by N_{11}^+, are symmetric. This property differs from the six relations in the lower half of the graph (Figure 3b), denoted by N_{11}^-, where the vertical axis forms a symmetry axis and corresponding relations form pairs of converse relations (A *inside* $B \Leftrightarrow B$ *contains* A and A *covers* $B \Leftrightarrow B$ *coveredBy* A). Elements that are located on the symmetry axis are symmetric. *Overlap* is part of N_{11}^+ and N_{11}^-; therefore, it fulfills the properties of both sets of relations. *Overlap* is also referred to as the *center element* of N_{11}. These properties would not change if one considers the additional connections that apply to identical region sizes or isotropic scalings that change boundaries at all points simultaneously. To account for such transitions, the neighborhood graph would add a vertical link from *equal* through *overlap* to *attaches* (for two identical half spheres) and two horizontal links—one from *inside* through *equal* to *contains*, and another one from *embraces* through *attaches* to *disjoint* (for isotropic scalings).

Table 1. The topological distance (Egenhofer and Al-Taha 1992) between the eleven topological relations in $\mathbb{IP}^2$. Highlighted is the shortest distance of the paths from the target relation (vertical) to the reference relation (horizontal), defining conceptual neighbors

$\tau(r_a, r_b)$		d	m	o	cb	cv	i	ct	e	a	en	em
d	$\begin{pmatrix} \varnothing & \varnothing & \neg\varnothing \\ \varnothing & \varnothing & \neg\varnothing \\ \neg\varnothing & \neg\varnothing & \neg\varnothing \end{pmatrix}$	0	1	4	5	5	6	6	6	4	7	6
m	$\begin{pmatrix} \varnothing & \varnothing & \neg\varnothing \\ \varnothing & \neg\varnothing & \neg\varnothing \\ \neg\varnothing & \neg\varnothing & \neg\varnothing \end{pmatrix}$	1	0	3	4	4	5	5	5	3	6	7
o	$\begin{pmatrix} \neg\varnothing & \neg\varnothing & \neg\varnothing \\ \neg\varnothing & \neg\varnothing & \neg\varnothing \\ \neg\varnothing & \neg\varnothing & \neg\varnothing \end{pmatrix}$	4	3	0	3	3	4	4	6	6	3	4
cb	$\begin{pmatrix} \neg\varnothing & \varnothing & \varnothing \\ \neg\varnothing & \neg\varnothing & \varnothing \\ \neg\varnothing & \neg\varnothing & \neg\varnothing \end{pmatrix}$	5	4	3	0	5	1	6	3	5	4	5
cv	$\begin{pmatrix} \neg\varnothing & \neg\varnothing & \neg\varnothing \\ \varnothing & \neg\varnothing & \neg\varnothing \\ \varnothing & \varnothing & \neg\varnothing \end{pmatrix}$	5	4	3	5	0	7	1	3	5	4	5
i	$\begin{pmatrix} \neg\varnothing & \varnothing & \varnothing \\ \neg\varnothing & \varnothing & \varnothing \\ \neg\varnothing & \neg\varnothing & \neg\varnothing \end{pmatrix}$	6	5	4	1	7	0	6	4	6	5	4
ct	$\begin{pmatrix} \neg\varnothing & \neg\varnothing & \neg\varnothing \\ \varnothing & \varnothing & \neg\varnothing \\ \varnothing & \varnothing & \neg\varnothing \end{pmatrix}$	6	5	4	6	1	6	0	4	6	5	4
e	$\begin{pmatrix} \neg\varnothing & \varnothing & \varnothing \\ \varnothing & \neg\varnothing & \varnothing \\ \varnothing & \varnothing & \neg\varnothing \end{pmatrix}$	6	5	6	3	3	4	4	0	4	5	6
a	$\begin{pmatrix} \varnothing & \varnothing & \neg\varnothing \\ \varnothing & \neg\varnothing & \varnothing \\ \neg\varnothing & \varnothing & \varnothing \end{pmatrix}$	4	3	6	5	5	6	6	4	0	3	4
en	$\begin{pmatrix} \neg\varnothing & \neg\varnothing & \neg\varnothing \\ \neg\varnothing & \neg\varnothing & \varnothing \\ \neg\varnothing & \varnothing & \varnothing \end{pmatrix}$	7	6	3	4	4	5	5	5	3	0	1
em	$\begin{pmatrix} \neg\varnothing & \neg\varnothing & \neg\varnothing \\ \neg\varnothing & \varnothing & \varnothing \\ \neg\varnothing & \varnothing & \varnothing \end{pmatrix}$	6	7	4	5	5	4	4	6	4	1	0

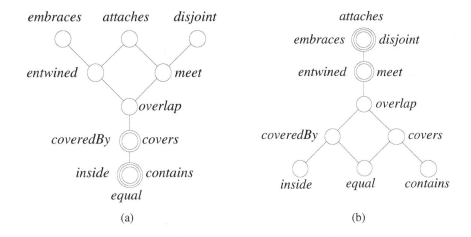

Fig. 3. Two orthogonal views of the conceptual neighborhood graph of the eleven spherical topological relations highlighting (a) the upper half and (b) the lower half

The following interpretations in terms of gradual movements can be made about the conceptual neighborhoods of the three exclusively spherical relations:

- Starting with the topological relation *meet*, where the boundaries partially intersect ($\partial A \cap \partial B = \neg\varnothing$ and $\partial A \cap B^- = \neg\varnothing$ and $A^- \cap \partial B = \neg\varnothing$) such that the two objects do not share any interior ($A° \cap B° = \varnothing$). If the two objects are gradually transformed such that they share more and more of their boundaries, without having their interiors intersect, then the relation *meet* will change into *attach* the moment the two boundaries coincide ($\partial A \cap \partial B = \neg\varnothing$ and $\partial A \cap B^- = \varnothing$ and $A^- \cap \partial B = \varnothing$); therefore, *meet* and *attach* are conceptual neighbors.
- Starting with the topological relation *attach* ($\partial A \cap \partial B = \neg\varnothing$ and $\partial A \cap B° = \varnothing$ and $A° \cap \partial B = \varnothing$). By pushing part of the boundary of one object into the interior of the other ($\partial A \cap \partial B = \neg\varnothing$ and $\partial A \cap B° = \neg\varnothing$ and $A° \cap \partial B = \neg\varnothing$), the two objects are *entwined*. A similar transition is possible from *overlap* to *entwined* by moving the entire part of the boundary that is located in the other object's exterior ($\partial A \cap B^- = \neg\varnothing$, $A^- \cap \partial B = \neg\varnothing$, and $\partial A \cap \partial B = \neg\varnothing$) from the exterior into the boundary ($\partial A \cap B^- = \varnothing$, $A^- \cap \partial B = \varnothing$, and $\partial A \cap \partial B = \neg\varnothing$) while maintaining a non-empty interior-interior intersection ($A° \cap B° = \neg\varnothing$). Since both of these transformations can be performed without the need of going through a third relation, *entwined* is a neighbor of both *attach* and *overlap*.
- Starting with the topological relation *entwined*, where part of the boundary is located in the other object's interior ($\partial A \cap B° = \neg\varnothing$ and $A° \cap \partial B = \neg\varnothing$), the remainder intersects with the other object's boundary ($\partial A \cap \partial B = \neg\varnothing$, $\partial A \cap B^- = \varnothing$, and $A^- \cap \partial B = \varnothing$). If the part of the boundary that intersects with the other object's boundary is moved completely into the other object's interior such that all of A's boundary is located in B's interior ($\partial A \cap B° = \neg\varnothing$, $A° \cap \partial B = \neg\varnothing$, and $\partial A \cap \partial B = \varnothing$), then A *embraces* B.

For display reasons, we employ a flattened graph, in which N_{11}^+ has been rotated by 90° such that all eleven relations fall into the same plane (Figure 4). This diagram also highlights the role of *overlap* as the center element of N_{11}. Relations are *at the same level* if they are located in the same part of the neighborhood graph (i.e., the upper half or the lower half) and if they have the same shortest path length from *overlap*. For instance, *inside*, *equal*, and *contains* are at the same level, because they are all in the lower half and the length of their shortest paths from *overlap* is 2.

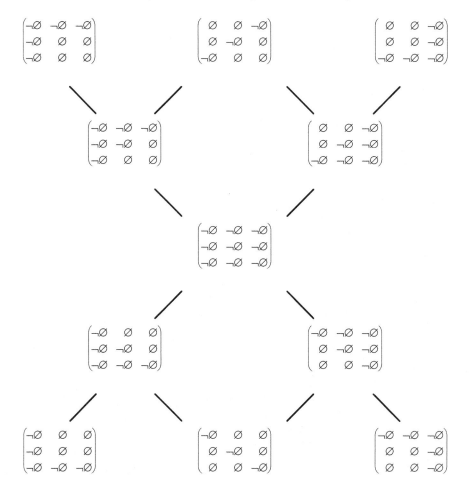

Fig. 4. The flattened conceptual neighborhood graph of the spherical topological relations

Considering the 9-intersection matrices within the organization of the conceptual neighborhood graph, common properties of a partially ordered set (Birkhoff 1967) are found:

- The least upper bound of any two topological relations at the same level is the intersection of the two relations' 9-intersection matrices.
- The greatest lower bound of any two topological relations at the same level is the union of the two relations' 9-intersection matrices.

The neighborhood graph also shows other regularities about the distribution of the elements in the 9-intersection matrices I (Equation 7).

$$I = \begin{pmatrix} i_{00} & i_{10} & i_{20} \\ i_{01} & i_{11} & i_{21} \\ i_{02} & i_{12} & i_{22} \end{pmatrix} \tag{7}$$

- Under the transposition along the horizontal axis through N_{11}'s center element, corresponding 9-intersection matrices, denoted by I^{T-}, are *horizontal* mirror images of each other (Equation 8).

$$\forall I \in N_{11}: \quad I^{T-} = \begin{pmatrix} i_{02} & i_{12} & i_{22} \\ i_{01} & i_{11} & i_{21} \\ i_{00} & i_{10} & i_{20} \end{pmatrix} \tag{8}$$

- Under the transposition along the vertical axis through N_{11}'s center element, corresponding 9-intersection matrices, denoted by $I^{T|}$, are mirror images along the *minor diagonal* (from top right to bottom left) for all intersections in N_{11}^+ (Equation 9a). The same property applies to all mirror images along the *main diagonal* (from top left to bottom right) for all intersections in N_{11}^- (Equation 9b).

$$\forall I \in N_{11}^+: \quad I^{T|} = \begin{pmatrix} i_{22} & i_{21} & i_{20} \\ i_{12} & i_{11} & i_{10} \\ i_{02} & i_{01} & i_{00} \end{pmatrix} \tag{9a}$$

$$\forall I \in N_{11}^-: \quad I^{T|} = \begin{pmatrix} i_{00} & i_{01} & i_{02} \\ i_{10} & i_{11} & i_{12} \\ i_{20} & i_{21} & i_{22} \end{pmatrix} \tag{9b}$$

- Under the transposition along N_{11}'s main diagonal, corresponding 9-intersection matrices, denoted by $I^{T/}$, are *vertical* mirror images of each other (Equation 10).

$$\forall I \in N_{11}: \quad I^{T/} = \begin{pmatrix} i_{20} & i_{10} & i_{00} \\ i_{21} & i_{11} & i_{21} \\ i_{22} & i_{12} & i_{02} \end{pmatrix} \tag{10}$$

- Finally, under the transposition along N_{11}'s minor diagonal, corresponding 9-intersection matrices, denoted by $I^{T\backslash}$, are also *vertical* mirror images of each other (Equation 11).

$$\forall I \in N_{11}: \quad I^{T\backslash} = \begin{pmatrix} i_{20} & i_{10} & i_{00} \\ i_{21} & i_{11} & i_{21} \\ i_{22} & i_{12} & i_{02} \end{pmatrix} \tag{11}$$

With these insights about the conceptual neighborhood graph of the topological relations in $\mathbb{P}^2$ we can answer the third question.

- The conceptual neighborhoods of all relations in $\mathbb{P}^2$ provide a consistent and regular framework for organizing the binary topological relations according to their similarity.

6   Inferences About Topological Relations in $\mathbb{P}^2$

The relations derived in the previous sections allow us to process topological queries on the sphere in a consistent fashion, but these relations *per se* do not allow us to perform any higher-level inferences about combinations of the relations. Such combinations are of interest if a query response cannot be derived directly from the stored base relations (Egenhofer and Sharma 1993). They are also relevant to assess whether a more complex query of conjunctions of such relations can produce a result at all or whether it is internally inconsistent (Egenhofer 1994b). The latter is also useful for assessing formally whether two or more independently collected sets of spatial descriptions conform or whether they contradict each other.

6.1   Single-Relation Inferences in $\mathbb{P}^2$

Some basic inferences over single relations can be made simply based on the properties of the conceptual neighborhood graph N_{11} and the relations' 9-intersection matrices. Among the eleven spherical relations we find two pairs of converse relations (Equations 12a-b), while each of the remaining seven relations is symmetric (Equation 12c-i).

$$inside\ (A,B) \Leftrightarrow contains\ (B,A) \tag{12a}$$

$$covers\ (A,B) \Leftrightarrow coveredBy\ (B,A) \tag{12b}$$

$$disjoint\ (A,B) \Leftrightarrow disjoint\ (B,A) \tag{12c}$$

$$meet\ (A,B) \Leftrightarrow meet\ (B,A) \tag{12d}$$

$$overlap\ (A,B) \Leftrightarrow overlap\ (B,A) \tag{12e}$$

$$equal\ (A,B) \Leftrightarrow equal\ (B,A) \tag{12f}$$

$$attaches\ (A,B) \Leftrightarrow attaches\ (B,A) \tag{12g}$$

$$entwined\ (A,B) \Leftrightarrow entwined\ (B,A) \tag{12h}$$

$$embraces\ (A,B) \Leftrightarrow embraces\ (B,A) \tag{12i}$$

6.2   Composition Table in $\mathbb{P}^2$

The basis for inferences over multiple relations is the composition (Tarski 1941). For a pair of spatial relations A r_i B and B r_j C, it determines the relation (or set of relations) that may hold between A and C. Typically composition of two relations is written as r_i ; r_j, omitting the references to the objects involved. For a set of n relations, the *composition table* captures all n^2 compositions. Subsequently we derive the composition table for the eleven topological relations in $\mathbb{P}^2$ and compare their inference power with that of the eight topological relations in $\mathbb{R}^2$.

To display the result of compositions in a compact format, we employ an iconic representation, in which each icon is based on the conceptual neighborhood graph (Figure 4). If a relation is part of the composition, the icon highlights it in the graph (Figure 5a). An icon with more than one highlighted relation implies that the composition results in multiple alternatives (Figure 5b). If all relations are highlighted, the composition of those particular relations yields the universal relation, which does not provide any inference information (Figure 5c).

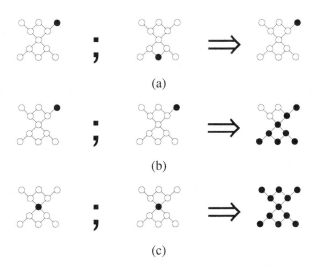

Fig. 5. Iconic presentations of compositions: (a) with a unique result, (b) with alternatives, and (c) with the universal relation as the result

We developed systematically the compositions of the spherical topological relations using the same method as for the composition of the topological relations in $\mathbb{R}^2$ (Egenhofer 1994a).

- A non-empty intersection between two parts A and B implies a non-empty intersection between the parts A and C if B is a subset of C (Equation 13a).
- An empty intersection between the parts A and B implies an empty intersection between the parts A and C if C is a subset of B (Equation 13b).

- A non-empty intersection between the parts A and B implies a non-empty intersection with the union of the two parts C_0 and C_1 if B is a subset of the union of C_0 and C_1 such that B intersects with both C_0 and C_1 (Equation 13c).
- An empty intersection between A and the union of B_0 and B_1 implies an empty intersection between A and C if C is a subset of the union of B_0 and B_1 (Equation 13d).

$$A \cap B = \neg \varnothing \ \wedge \ B \subseteq C \ \Rightarrow \ A \cap C = \neg \varnothing \tag{13a}$$

$$A \cap B = \varnothing \ \wedge \ B \supseteq C \ \Rightarrow \ A \cap C = \varnothing \tag{13b}$$

$$A \cap B = \neg \varnothing \ \wedge \ B \subseteq (C_0 \cup C_1) \ \wedge \ B \not\subseteq C_0 \ \wedge \ B \not\subseteq C_1 \tag{13c}$$
$$\Rightarrow \ A \cap C_0 = \neg \varnothing \ \wedge \ A \cap C_1 = \neg \varnothing$$

$$A \cap (B_0 \cup B_1) = \varnothing \ \wedge \ (B_0 \cup B_1) \supseteq C \ \wedge \ B_0 \not\supseteq C \ \wedge \ B_1 \not\supseteq C \tag{13d}$$
$$\Rightarrow \ A \cap C = \varnothing$$

The composition of all 121 pairs of spherical topological relations was determined computationally with a Prolog program with a total of 44 lines of code (11 ground axioms for the 11 base relations and 33 predicates to determine the inferred compositions). It ran 6.5 seconds on a 266 MHz Macintosh PowerBook G3. Figure 6 displays the 11×11 composition table for the topological relations that can be realized on the sphere between two regions.

A comparison of the counts of relations in each composition reveals interesting similarities among the eight planar relations and the three exclusively spherical relations:

- All compositions with *equal* have the same cardinality (i.e., number of relations) as the compositions with *attach*. One interpretation is that the coincidence of the boundaries, which is common to both relations, is a strong factor for making inferences.
- All compositions with *coveredBy* have the same cardinality as the compositions with *entwined*. This analogy has the same roots as the matching between *equal* and *attach*.
- All compositions with *inside* have the same cardinality as the compositions with *embrace*. In both cases, one region's boundary is completely contained in the interior of the other region's boundary.

The composition table is the foundation for assessing whether or not the spherical topological relations form a relation algebra (Tarski 1941). Using the set-theoretic operations union ($\cup$), intersection ($\cap$), and complement ($-$), and considering *equal* as the identity relation and $\bar{r}$ as the converse relation of r (Equation 12a-i), we found that all seven properties of an relation algebra are fulfilled by the set of eleven spherical topological relations:

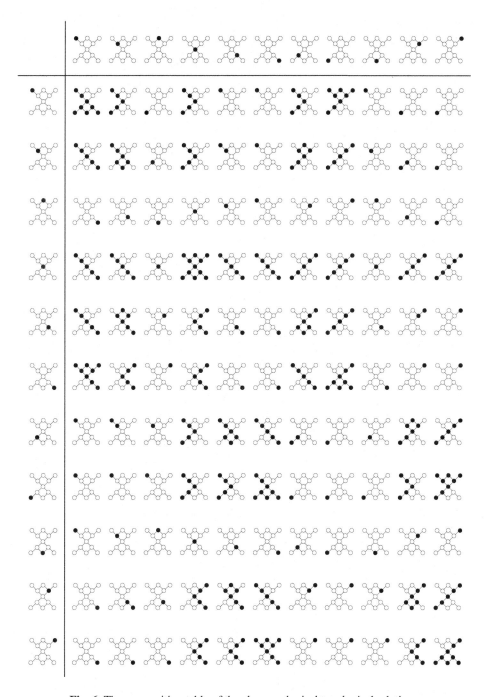

Fig. 6. The composition table of the eleven spherical topological relations

- Each composition with the identity relation is idempotent, because $\forall r:\ r\,;equal = r$.
- The composition with a set of relations is equal to the union of the compositions with each of the elements of the set, because $\forall (r_i, r_k)\exists r_j:\ (r_i \cup r_j)\,;r_k = (r_i\,;r_k)\cup(r_j\,;r_k)$.
- The converse of a converse relation is equal to the original relation, because $\forall r:\ \overline{\overline{r}} = r$.
- The converse of a set of relations is equal to the union of the converse relations of each of the elements of that set, because $\forall (r_i, r_j):\ \overline{(r_i \cup r_j)} = \overline{r_i}\cup\overline{r_j}$.
- The converse relation of a composition is equal to the composition of the converses of the two relations, taken in reverse order, because $\forall (r_i, r_j):\ \overline{(r_i\,;r_j)} = \overline{r_j}\,;\overline{r_i}$.
- A variation of De Morgan's Theorem K holds, because $\forall (r_i, r_j):\ \overline{r_i}\,; -(r_i\,;r_j)\cup - r_j = - r_j$.

The composition is associative, because $\forall (r_i, r_k)\exists r_j:\ (r_i\,;r_j)\,;r_k = r_i\,;(r_j\,;r_k)$.

6.3   Comparing the Inference Power of Topological Relations in $\mathbb{R}^2$ and $\mathbb{P}^2$

It was expected that spherical topological reasoning would be more complex than topological reasoning in $\mathbb{R}^2$. If this assumption was true then the composition tables for $\mathbb{R}^2$ and $\mathbb{P}^2$ should reveal that the addition of the three spherical relations makes the inferencing less crisp.

To assess the crispness of compositions, we use four different measures. First we count the number of relations in a composition (Equation 14). The more relations in a composition, the less crisp the inference and, therefore, the more possible cases a person or a machine needs to consider as the outcome of an inference.

$$C = \#(r_i; r_j) \tag{14}$$

Since the largest number differs for relations in $\mathbb{R}^2$ and $\mathbb{P}^2$ (it is 8 vs. 11), we use a second crispness measure $\overline{C}$, a normalized count of relations (Equation 15). It has a different base for $\mathbb{R}^2$ and $\mathbb{P}^2$ (i.e., 8 and 11, respectively). This normalized crispness measure has the highest value if the composition is unique, while it is 0 for a composition that results in the universal relation.

$$\overline{C}_8 = 1 - \frac{\#(r_i; r_j)}{8} \tag{15a}$$

$$\overline{C}_{11} = 1 - \frac{\#(r_i; r_j)}{11} \tag{15b}$$

The last two measures are based on the number of undetermined compositions (i.e., compositions that result in the universal relation, which does not yield any inferences at all) and the number of determined compositions (i.e., compositions that result in a single relation, which allows for the most crisp inference). They indicate how often nothing can be derived, or how often the inference is unique.

Hypothesis 1:   The composition of $\mathbb{P}^2$-topological relations is less crisp than the composition of $\mathbb{R}^2$-topological relations, because it has more undetermined compositions.

Dismissed. The $\mathbb{R}^2$-composition table has three universal relations—the results of (1) *disjoint* ; *disjoint*, (2) *overlap* ; *overlap*, and (3) *inside* ; *contains*—while the $\mathbb{P}^2$-composition table has a single universal relation (the result of *overlap* ; *overlap*). Normalized over the total number of compositions, this means a decrease in undetermined compositions from 4.7% to 0.8%. None of the compositions with any of the three exclusively spherical relations is undetermined, and the least crisp compositions involving the exclusively spherical relations is 8 out of 11 (i.e., $\overline{C}_{11} = 0.273$), which occurs for three compositions—(1) embrace ; contains, (2) inside ; embrace, and (3) embrace ; embrace.   ∴

Hypothesis 2:   The composition of $\mathbb{P}^2$-topological relations is less crisp than the composition of $\mathbb{R}^2$-topological relations, because it has fewer determined compositions.

Dismissed. The $\mathbb{R}^2$-composition table has 27 compositions of cardinality 1, while the $\mathbb{P}^2$-composition table has 64 of such crisp compositions. Normalized over the total number of compositions, this means an increase in determined compositions from 42.1% to 52.9%. The relations with the highest numbers of determined compositions in $\mathbb{P}^2$ are *equal* and *attach* (all compositions with *equal* and *attach* are determined), while the lowest number of determined compositions involving a particular relation is with *overlap*.   ∴

Hypothesis 3:   The composition of all eleven spherical topological relations is less crisp than the composition of the eight topological relations in $\mathbb{R}^2$.

Dismissed for relative counts, but confirmed for absolute counts. The average crispness of all 64 $\mathbb{R}^2$-compositions is $\overline{C}_8 = 0.623$, while $\overline{C}_{11} = 0.727$ for all 121 $\mathbb{P}^2$-compositions. The crispness of the 57 compositions that involve at least one exclusively spherical relation is also higher ($\overline{C}_{11} = 0.781$) than the average of all $\mathbb{R}^2$-compositions ($\overline{C}_8 = 0.623$). In absolute numbers, all compositions in $\mathbb{R}^2$ include 193 relations, while there are 363 compositions in $\mathbb{P}^2$. When compared with respect to the total number of compositions existing in $\mathbb{R}^2$ and $\mathbb{P}^2$, the two ratios are almost identical: on average there are 3.01 relations per composition in $\mathbb{R}^2$ and 3.00 in $\mathbb{P}^2$.   ∴

Hypothesis 4:   The addition of the three exclusively spherical relations reduces the crispness of the majority of the 64 compositions.

Dismissed. When projecting the eight $\mathbb{R}^2$-relations onto $\mathbb{P}^2$, 13 of their 64 compositions become less crisp (for these 13 relations, the crispness $\overline{C}_8$ is by an average of 7.5% greater than $\overline{C}_{11}$). Fifty of the 64 compositions become crisper (for these 50 relations, the crispness $\overline{C}_8$ is by an average of 9.1% smaller than $\overline{C}_{11}$). One composition—*overlap* ; *overlap*—is equally crisp in $\mathbb{R}^2$ and in $\mathbb{P}^2$ (i.e., it has the same inference power in $\mathbb{R}^2$ as in $\mathbb{P}^2$). For all 64 compositions, this means an average increase in crispness by 5.6%.

While in absolute numbers the total count goes up from 193 in $\mathbb{R}^2$ to 226 in $\mathbb{P}^2$ (i.e., a 17% increase), the loss of crispness comes from 15 out of the 64 compositions (23.4%), while the remainder (76.6%) stays unchanged. Among the 15 compositions that loose inference crispness by mapping the relations from $\mathbb{R}^2$ onto $\mathbb{P}^2$, eight decrease in their crispness by 67%, two by 60%, three by 50%, one by 38%, and another one by 33%. The greatest decrease in crispness occurs for compositions that involve *overlap*, while—as expected—all compositions with *equal* remain perfectly crisp.   ∴

Hypothesis 5:   Compositions of topological relations on the sphere are more often undetermined than compositions of topological relations in the plane.

Dismissed. In $\mathbb{P}^2$ there is only one undetermined composition (i.e., 0.8% of all $\mathbb{P}^2$-compositions), whereas in $\mathbb{R}^2$ there are three undetermined compositions (which corresponds to 4.7% of all possible compositions between $\mathbb{R}^2$-relations).   ∴

Hypothesis 6:   Compositions of topological relations on the sphere are less often uniquely determined than compositions of topological relations in the plane.

Dismissed. In $\mathbb{P}^2$ there are 64 unique compositions (which is 52.9% of all 121 $\mathbb{P}^2$-compositions), while in $\mathbb{R}^2$ there are 27 unique compositions (i.e., 42.1% of the 64 compositions between $\mathbb{R}^2$-relations).   ∴

7   Conclusions

Models of geographic space as they are used in current geographic information systems are typically oversimplified. They reduce, for instance, the 3-dimensional nature of geographic phenomena to a planar view, and they flatten the surface of the Earth from a sphere into the plane. Such simplifications are sufficiently good approximations for capturing locally limited geographic areas, but impose serious limitations for modeling global geographic phenomena, as required in such a setting as Digital Earth.

This paper investigated fundamental spatial properties that are preserved in the transition from a flat, two-dimensional embedding space to a two-dimensional surface embedded in a three-dimensional space. Such a setting corresponds to modeling and analyzing spatial phenomena on the surface of a sphere. With a focus on topological relations, we gained new insights about qualitative topological reasoning, comparing the planar with the spherical setting. While the sphere offers additional topological relations that cannot be realized in the plane—the set of possible relations grows by 37.5% from 8 to 11—the inferences that can be made with the composition of topological relations remain primarily crisp, which is a measure of the inference power of the algebraic system formed by these topological relations. Only 23.4% compositions are diluted if the setting for their analysis is on the sphere rather than in the plane. On the other hand, compositions that involve at least one relation that can only be realized on the sphere are on average crisper than compositions of relations that can be realized in the plane. Based on these analyses we conclude that the transition from planar to spherical topological reasoning is a small step that should require few additional logical reasoning abilities.

The second insight relates to the parallel between one-dimensional (e.g., temporal) reasoning and two-dimensional (e.g., spatial) reasoning. We found that the transition from the plane to the sphere (for the two-dimensional case) corresponds to the transition from a linear model to a cyclic model (for the one-dimensional case). This finding is based on the observation that both transitions give rise to additional qualitative relations. These additional relations extend the conceptual neighborhood graphs in parallel ways (even though the linear relations used in $\mathbb{R}^1$ are based on an orientation of the plane and, therefore, typically create pairs of converse relations where there is only one orientation-neutral relation in the two-dimensional setting). Such analogies are critical to increase our understanding about the relationship between spatial and temporal reasoning, in particular providing answers to why certain types of spatial and temporal concepts appear to be compatible.

A final observation relates to the stunning regularity of the numbers of unique compositions in $\mathbb{R}^2$ ($27=3^3$) and $\mathbb{P}^2$ ($64=4^3$).

Acknowledgments

Max Egenhofer's research is partially supported by the National Science Foundation under NSF grants IIS-9970123 and EPS-9983432; the National Geospatial-Intelligence Agency under grant numbers NMA202-97-1-1023, NMA201-00-1-2009, NMA201-01-1-2003, and NMA401-02-1-2009; and the National Institute of Environmental Health Sciences, NIH, under grant number 1 R 01 ES09816-01. An earlier version of this paper was presented at *The "I" in GIScience: Fundamental Questions about the Nature of Geographic Information*, held in Manchester, UK, July 2001. I am grateful to all participants who provided feedback at that meeting.

References

1. J. Allen (1983) Maintaining Knowledge about Temporal Intervals. *Communications of the ACM* 26(11): 832-843.
2. P. Balbiani and A. Osmani (2000) A Model for Reasoning about Topological Relations between Cyclic Intervals. in: A. Cohn, F. Giunchiglia, and B. Selman (Eds.) *Seventh International Conference on Principles of Knowledge Representation and Reasoning, KR2000*, Breckenridge, CO, pp. 675-687, San Mateo, Morgan Kaufmann Publishers.
3. R. Billen, S. Zlatanova, P. Mathonet, and F. Bouvier (2002) The Dimensional Model: A Framework to Distinguish Spatial Relationships. in: D. Richardson and P. van Oosterom (Eds.) *Advances in Spatial Data Handling: Tenth International Symposium on Spatial Data Handling*: 285-298, Berlin, Springer.
4. G. Birkhoff (1967) *Lattice Theory*. Providence, RI, American Mathematical Society.
5. T. Bruns and M. Egenhofer (1996) Similarity of Spatial Scenes. in: M.-J. Kraak and M. Molenaar (Eds.) *Seventh International Symposium on Spatial Data Handling*, Delft, The Netherlands, pp. 173-184, London, Taylor & Francis.
6. E. Clementini and P. di Felice (1997) Approximate Topological Relations. *International Journal of Approximate Reasoning* 16(2): 173-204.
7. E. Clementini, P. di Felice, and G. Califano (1995) Composite Regions in Topological Queries. *Information Systems* 20(7): 579-594.
8. E. Clementini, P. di Felice, and P. van Oosterom (1993) A Small Set of Formal Topological Relationships Suitable for End-User Interaction. in: D. Abel and B. C. Ooi (Eds.) *Third International Symposium on Large Spatial Databases, SSD '93*. Lecture Notes in Computer Science 692: 277-295. New York, NY, Springer-Verlag.

9. E. Clementini, J. Sharma, and M. Egenhofer (1994) Modelling Topological Spatial Relations: Strategies for Query Processing. *Computers and Graphics* 18(6): 815-822.

10. A. Cohn and N. Gotts (1996) The "Egg-Yolk" Representation of Regions with Indeterminate Boundaries. in: P. Burrough and A. Frank (Eds.) *Geographic Objects with Indeterminate Boundaries*: 171-187. London, Taylor & Francis.

11. Z. Cui, A. Cohn, and D. Randell (1993) Qualitative and Topological Relationships in Spatial Databases. in: D. Abel and B. Ooi (Eds.) *Third International Symposium on Large Spatial Databases*. Lecture Notes in Computer Science 692: 296-315. New York, NY, Springer-Verlag.

12. M. Egenhofer (1994a) Deriving the Composition of Binary Topological Relations. *Journal of Visual Languages and Computing* 5(2): 133-149.

13. M. Egenhofer (1994b) Pre-Processing Queries with Spatial Constraints. *Photogrammetric Engineering & Remote Sensing* 60(6): 783-790.

14. M. Egenhofer and K. Al-Taha (1992) Reasoning About Gradual Changes of Topological Relationships. in: A. Frank, I. Campari, and U. Formentini (Eds.) *Theories and Methods of Spatio-Temporal Reasoning in Geographic Space, Pisa, Italy*. Lecture Notes in Computer Science 639: 196-219. Berlin, Springer-Verlag.

15. M. Egenhofer and R. Franzosa (1991) Point-Set Topological Spatial Relations. *International Journal of Geographical Information Systems* 5(2): 161-174.

16. M. Egenhofer and R. Franzosa (1995) On the Equivalence of Topological Relations. *International Journal of Geographical Information Systems* 9(2): 133-152.

17. M. Egenhofer and J. Herring (1991) Categorizing Binary Topological Relationships Between Regions, Lines, and Points in Geographic Databases. in: M. Egenhofer, J. Herring, T. Smith, and K. Park (Eds.) *A Framework for the Definition of Topological Relationships and an Algebraic Approach to Spatial Reasoning within this Framework, NCGIA Technical Report 91-7*. Santa Barbara, CA, National Center for Geographic Information and Analysis.

18. M. Egenhofer and D. Mark (1995) Modeling Conceptual Neighbourhoods of Topological Line-Region Relations. *International Journal of Geographical Information Systems* 9(5): 555-565.

19. M. Egenhofer, P. Di Felice, and E. Clementini (1994) Topological Relations between Regions with Holes. *International Journal of Geographical Information Systems* 8(2): 129-144.

20. M. Egenhofer and J. Sharma (1993) Assessing the Consistency of Complete and Incomplete Topological Information. *Geographical Systems* 1(1): 47-68.

21. M. Egenhofer, J. Sharma, and D. Mark (1993) A Critical Comparison of the 4-Intersection and 9-Intersection Models for Spatial Relations: Formal Analysis. in: R. McMaster and M. Armstrong (Eds.) *Autocarto 11*, Minneapolis, MN, pp. 1-11.

22. C. Freksa (1992) Temporal Reasoning Based on Semi-Intervals. *Artificial Intelligence* 54: 199-227.

23. N. Gott (1996) *Using the 'RCC' Formalism to Describe the Topology of Spherical Regions*. Technical Report 96.24, Leeds, University of Leeds.

24. T. Hadzilacos and N. Tryfona (1992) A Model for Expressing Topological Integrity Constraints in Geographic Databases. in: A. Frank, I. Campari, and U. Formentini (Eds.) *Theories and Methods of Spatio-Temporal Reasoning in Geographic Space*. Lecture Notes in Computer Science 639: 252-268. Pisa, Springer-Verlag.

25. N. W. Hazelton, L. Bennett, and J. Masel (1992) Topological Structures for 4-Dimensional Geographic Information Systems. *Computers, Environment, and Urban Systems* 16(3): 227-237.

26. D. Hernández (1994) *Qualitative Representation of Spatial Knowledge.* New York, Springer-Verlag.
27. K. Hornsby, M. Egenhofer, and P. Hayes (1999) Modeling Cyclic Change. in: P. Chen, D. Embley, J. Kouloumdjian, S. Liddle, and J. Roddick (Eds.) *Advances in Conceptual Modeling, Versailles, France.* Lecture Notes in Computer Science 1227: 98-109. Berlin, Springer-Verlag.
28. IBM (2002) *IBM Informix Geodetic DataBlade Module.* User's Guide, Version 3.11, available from http://publib.boulder.ibm.com/epubs/pdf/8675.pdf, White Planes, NY, IBM Corporation.
29. D. Mark and M. Egenhofer (1994) Modeling Spatial Relations Between Lines and Regions: Combining Formal Mathematical Models and Human Subjects Testing. *Cartography and Geographic Information Systems* 21(3): 195-212.
30. D. Papadias, Y. Theodoridis, T. Sellis, and M. Egenhofer (1995) Topological Relations in the World of Minimum Bounding Rectangles: A Study with R-Trees. *ACM SIGMOD* 4(2): 92-103.
31. C. Papadimitriou, D. Suciu, and V. Vianu (1996) Topological Queries in Spatial Databases, *Fifteenth ACM SIGACT-SIGMOD-SIGART Symposium on Principles of Database Systems (PODS)*, Montreal, Canada, pp. 81-92, ACM Press.
32. S. Pigot (1991) Topological Models for 3D Spatial Information Systems. in: D. Mark and D. White (Eds.) *Autocarto 10*, Baltimore, MD, pp. 368-392.
33. D. Pullar and M. Egenhofer (1988) Toward Formal Definitions of Topological relations Among Spatial Objects, in: *Third International Symposium on Spatial Data Handling*, Sydney, Australia, pp. 225-241.
34. D. Randell, Z. Cui, and A. Cohn (1992) A Spatial Logic Based on Regions and Connection. in: B. Nebel, C. Rich, and W. Swartout (Eds.) *Principles of Knowledge Representation and Reasoning, KR '92*, Cambridge, MA, pp. 165-176.
35. J. Sharma (1999) Integrated Topological and Directional Reasoning in Geographic Information Systems. in: M. Craglia and H. Onsrud (eds.), *Geographic Information Research: Trans-Atlantic Perspectives.* Taylor & Francis, London, pp. 435-447.
36. T. Smith and K. Park (1992) Algebraic Approach to Spatial Reasoning. *International Journal of Geographical Information Systems* 6(3): 177-192.
37. A. Tarski (1941) On the Calculus of Relations. *The Journal of Symbolic Logic* 6(3): 73-89.
38. L. Usery (2002) *University Consortium for Geographic Information Science Research Priorities: Global Representation and Modeling*, http://www.ucgis.org/priorities/research/2002researchPDF/shortterm/0_global_representation.pdf
39. S. Winter (1995) Topological Relations between Discrete Regions. in: M. Egenhofer and J. Herring (Eds.) *Advances in Spatial Databases—4th International Symposium, SSD '95, Portland, ME.* Lecture Notes in Computer Science 951: 310-327. Berlin, Springer-Verlag.

GeoPQL: A Geographical Pictorial Query Language That Resolves Ambiguities in Query Interpretation

Fernando Ferri[1] and Maurizio Rafanelli[2]

[1] IRPPS-CNR, via Nizza 128, 00187 Roma, Italy
f.ferri@irpps.cnr.it
[2] IASI-CNR, viale Manzoni 30, 00185 Roma, Italy
rafanelli@iasi.cnr.it

Abstract. The main problem of visual query languages for geographical data concerns the query's ambiguity. Ambiguity derives from the fact that a query can lead to multiple interpretations for both the system and user. In fact a query can have different visual representations, and these can themselves have different interpretations. Among the reasons leading to these ambiguities, one appears to be fundamental: the user gives his own semantics to the information. However his actions may not completely represent his intentions, so the system may make an incorrect interpretation. Additionally, when a user draws two icons representing different geographical objects of a query he cannot avoid defining one or more spatial relationships between them. This is the case for any pair of icons, however the user often does not want to define spatial relationships between all pair of icons. So he cannot express his exact query and different queries must be formulated to obtain his goals.

This work proposes a Pictorial Geographical Query Language, GeoPQL, that allows the user to represent only the desired relationships and avoid undesired relationships in the query's visual representation.

The language is based on twelve operators. The set of operators includes all the main topological operators, distance and two operators devoted to solving ambiguities in visual query representation. The paper then discusses syntactic and semantic correctness of spatial configurations and related operators in the context of the declarative geographic pictorial query language. Some possible ambiguities and their solutions are presented in order to show the language's characteristics.

GeoPQL has been implemented as a stand alone tool which interfaces with ESRI's ArcView®, and the main results obtained are: high expressive power, solution of the ambiguities inherent to the spatial representation of a query and exact matching between the query and the obtained results.

1 Introduction

There has recently been a great deal of research in the domain of geographical information systems (GIS). A fundamental research area concerns the definition of high level user interfaces, as one of the main characteristics of GIS is its management of complex and large amounts of data, while its users are generally non-computer scientists.

S. Spaccapietra and E. Zimányi (Eds.): Journal on Data Semantics III, LNCS 3534, pp. 50–80, 2005.
© Springer-Verlag Berlin Heidelberg 2005

For such users the proposed query languages (QL) are often very technical, so they can have great difficulty in formulating queries. Visual GIS query languages have the aim of solving this lack of user-friendliness. A query language is said to be visual (or graphical) whenever the query's semantics is expressed by a drawing. It is said to be declarative whenever the query specifies the properties to be verified by the result, but not the way of obtaining them. In the conceptual representation of geographical objects of the real world, the different GIS visual query language proposals consider only three types of symbolic graphical objects (sgo): point, polyline and polygon, to represent any geographic object and/or a limited set of graphical symbols in order to represent some operators that do not have a representation expressing the involved relation. Symbolic graphical objects (sgo) are usually referred to in literature by terms such as icons, symbols, feature, etc. Among the advantages of this type of query languages are ease of use, a natural logical approach (in the sense that the user applies operators without explicitly expressing them) and, sometimes, the fact that there is no need to know the language textual syntax. Such languages are particularly appropriate for Geographical Information Systems, due to the nature of their data.

In these kinds of languages a query can lead to multiple interpretations for the system and user. One of the main reasons is that a unique working space is often used to represent and express different kinds of information. Another is due to the different approach that the user has in formulating his query with respect to the analysis that the system makes of it. The fact that different interpretations are possible for the same query leads to the need to manage the ambiguity in its representation.

The proposal presented here stands out from other visual query languages due to its different characteristics. The main ones are: it solves ambiguities without the need for user-system dialog (for example as proposed in [1]) because the user, freely drawing his query, obtains a unique interpretation by use of the operators defined in GeoPQL; nothing is predefined (symbols, relationships between two symbols, etc) and syntactically correct relationships are considered by the system; the query language tool is implemented as a stand-alone query language, which interfaces with a commercial GIS.

The paper is structured as follows. Section 2 gives a brief overview of ambiguity in visual query languages for geographical data. Section 3 illustrates the best known previous literature proposals and discusses a query example, illustrating the differences between the user's intention (the so-called user mental model) and what is understood by each visual query language (ambiguity problem), and whether it is able to answer the query or not. Section 4 proposes the Geographical Pictorial Query Language (GeoPQL), discussing its operators in a geographical data context. In Section 5 the Syntactic and semantic correctness of the pictorial configurations is discussed. Section 6 shows how GeoPQL resolves eventual ambiguities that can arise in graphical representation of a query. Section 7 presents the implemented system and examples of pictorial queries. A conclusion is given in Section 8.

2 Ambiguity Related to Visual Queries for Geographical Data

Visual languages often use icons to model spatial objects, express spatial relationships between objects and, then, formulate queries. They offer an intuitive and incremental view of spatial queries, but often have little expressive power, fairly ineffective query execution and, may offer different interpretations of the same query. In particular, ambiguity is one of the main problems in using visual query languages for geographical data.

For example, suppose the user wants to formulate the following query: "Find all the regions which are *passed through* by a river and *overlap* a forest", where the term region means an Italian administrative subdivision (similarly to the German "lander" or the American state). In this query the user is not interested in the relationship between the river and the forest and the absence, in natural language formulation, of explicit relationships between them means that the phrase "irrespective of the topological relationship between the river and forest" should complete the query.

The different proposed visual query languages give a visual representation of the query "Find all the regions which are *passed through* by a river and *overlap* a forest" maintaining the ambiguity of the existing relationships between the river and the forest. So, any representation considering a specific relationship between the two sgo can be considered as valid. Different visual queries can thus represent the previous query in natural language. In particular, in Figure 1-a the forest and the river are "Disjointed", in Figure 1-b the river touches the forest and in Figure 1-c the river passes through the forest.

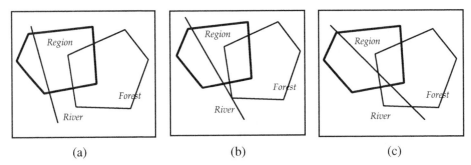

(a)	(b)	(c)

Fig. 1. Three visual queries representing the same query expressed in natural language

Moreover the user has to specify the target of the query highlighting the set of objects composing the target (in this case the object region). When a parser processes the queries in Figure 1, it should consider both the target and all relationships represented in the visual representation. For this reason, the three representations should be interpreted as three different queries with each having a different meaning to that of the original in natural language.

To remove the ambiguity, the complete natural language query "Find all the regions which are *passed through* by a river and *overlap* a forest, irrespectively of the topological relationships between the river and the forest" could be considered. However, when the user draws an sgo representing a forest and another representing a river he can not avoid representing a topological relationship between them. This means that the phrase "irrespectively of the topological relationship between the river and the forest" does not have a single, unique drawing: it is necessary to represent the logical OR of as many different drawings as there are valid configurations between the two sgo (shown in Figure 2). This is because in several approaches known in literature, the phrase "*any* configuration between two sgo is valid" has no corresponding drawing.

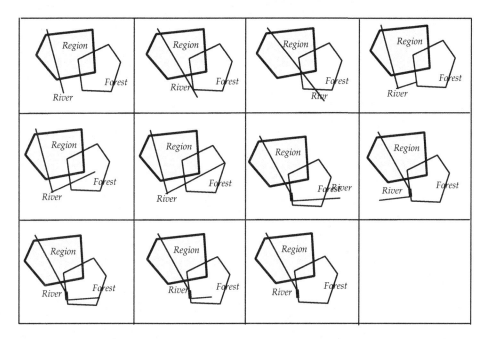

Fig. 2. Different drawings of valid configurations between River and Forest

3 Related Works

Several proposals of visual languages for geographical data exist in literature. These are discussed, using the same example query as in Figure 1, by illustrating if and how each language resolves the query, and showing the difference between the user's intention and what each system understands.

Some of these languages avoid multiple interpretations by considering only limited kinds of spatial relations, as in the Pictorial Query-by-Example (PQBE) [2], whose goal is to find directional relationships. The authors use the concept of symbolic image, which is an array representing a set of objects and a set of direction relationships among them: images could correspond to visual scenes, geographical maps or other forms of spatial data. A database consists of a set of symbolic images and a query consists of a skeleton image, which is itself a symbolic image. The main limitation of this approach is that the use of symbolic images represented by means array makes it difficult to represent distances and other types of non-directional spatial relationships. The language considers directional relationships only, and for this reason it is impossible to completely express the query in Figure 1, involving topological relationships. This language can express, for example, the query "find all the Regions which are north-west of a forest" represented by the skeleton image (the array) of Figure 3-a.

The SVIQUEL visual language [3] also includes topological operators considering 45 different types of primitives capable of representing topological and directional relationships between two sgo of type polygon. However SVIQUEL avoids multiple interpretations by limiting the number of objects involved (to just two) and providing a tool with a low expressive power for specifying the relative spatial positions (both topological and directional) based on the selection of one (or more) of the 45 primitives or the direct manipulation of 8 spatial filters called S-sliders (similar to scroll-bar) representing the reciprocal position of a pair of polygonal graphical objects. It is easy to use, but allows formulations of only very elementary queries, involving just two polygonal objects. For these reasons it is impossible to completely express the query in Figure 1, involving more than two sgo. The language can express, for example, the query "Find all the regions which *overlap* a forest" represented in Figure 3-b.

Other languages enlarge the user's possibilities both by allowing more complex queries to be formulated and by giving a free interaction modality. To date there have been two main approaches to designing these kinds of visual query languages: in the first the user draws icons and graphical symbols on the screen to express an operator or a geographical object, while in the second the user draws his query directly on the screen using the blackboard metaphor.

Several languages use the first approach, in particular Grog [4], a graph-oriented Data Manipulation Language for GIS applications, based on data represented using directed graphs. Manipulations of graphs are defined with a recursive or logic-based formalism. Queries are thus graphical and defined on a graph, as a set of labelled oriented graphs. The graph node labels can be variables or constants (they refer to objects of the geographical database). The graph's edges can be of three different types: link edge, inclusion edge and intersection edge. Each of them is oriented, represents the results of sub-queries and is a binary operator. They also represent paths without cycles. An answer is thus a set of graphs and each of its elements is a result graph. The final answer is a logical OR of the result graphs. This language therefore has only three operators and avoids ambiguities by using a graph to represent data which are not strictly graphs but are characterized to have a spatial dimension. This language can not express the query in Figure 1, because it considers only polyline-type sgo and is mainly aimed at representation of infrastructure networks such as railroads, roads, etc.

Another language, Cigales [5, 6], is based on two objects: line and area. Cigales allow the user to draw a query. It is based on the idea of expressing a query by drawing the pattern corresponding to the result desired by the user. To achieve this it uses a set of symbolic graphical objects, some of which, called graphic labels by the authors, are able to model the geometrical objects polyline and polygon (point is not considered), and the operations carried out on these objects (intersection, inclusion, adjacency, path and distance). Symbolic graphical objects and graphical labels conceptualising the operators are predefined. Each object is characterized by a set of attributes.

Lvis [7] is an extension of Cigales. The most relevant difference consists of the definition of new operators, as both spatial and temporal properties of the objects forming the query are considered. Four groups of operators are considered: Logical (And, Or and Not), Spatial (Intersection, Inclusion, Adjacency, Disjunction, Equality, Way, Distance, Radial distance and Zone), Temporal (Before, After, Overlapping, During, Equal, Beginning and End) and finally Spatio-temporal operators (Create, Destroy, Fusion, Increase, Reduce and Split). It is possible to carry out 4 types of queries: Thematic, Spatial, Temporal and Spatio-temporal. Every query is translated in the host language of the GIS.

The main limitation of these languages using the first approach is in obtaining different interpretations of the same query [1] [8], that is, the system is not able to give a unique interpretation of a given visual query representation. In [8] the authors affirm that both Grog, and Cigales "may produce misunderstandings between the end-users and the GIS query evaluator". The proposed merger of two graphical languages brings to two possible solutions: the former introduces various interactions (feed back) with the user, the latter increases the complexity of the resolution model. Moreover, different confusion cases may arise, so that "the semantics of the query are fully-user dependent". Finally, "complex query, with numerous basic objects are not expressible". Increasing the number of query objects also increases ambiguity. This ambiguity derives from the fact that a query can lead to multiple interpretations for both system and user. Among the reasons leading to these ambiguities, one appears fundamental: the user gives his own semantics to information, that is, he has the so called user's mental model. His actions may not represent his actual intentions, so the system may be led to an incorrect interpretation.

A possible visual query for Cigales and LVIS is represented in Figure 3-c. Obviously the query can be ambiguous and have a different interpretation than that intended by the user, such as "Find all the regions which are *passed through* by a river and *overlap* a forest, and in which the river is disjointed from the forest".

In an other approach [9] it is possible to remove undesired relationships among drawn symbolic graphical objects or impose an a priori restrictive interpretation using the foreground/background metaphor. When the user draws a new symbolic graphical object, he can set the state of all the previous drawn symbolic graphical objects to foreground or background. Symbolic graphical objects for which the relationships with the new symbolic graphical object have to be considered must be placed in the foreground, while those whose the relationships do not have to be considered are placed in the background.

The relationships of a new symbolic graphical object thus depend on the state (foreground or background) of the previously drawn symbolic graphical objects. A relationship between a newly drawn σ_i and a previous symbolic graphical object σ_j is considered if σ_j is in foreground state. It is not considered if σ_j is in background state. Using this approach, to interpret a query the parser must consider both the visual representation and the drawing process and more specifically the order in which the symbolic graphical objects are drawn and the state (foreground or background) of all the symbolic graphical objects when a new symbolic graphical object is drawn. In this manner some procedural steps influence the semantics of the query but they do not influence its representation, and queries having the same representation may have different semantics. The user check the result, translated into a textual form. He can thus avoid the generation of extraneous relationships.

The graphical representation of the query "Find all the regions which are *passed-through* by a river and *overlap a* forest" is shown in Figure 3-d, where the symbolic graphical object Forest must be in the background when the symbolic graphical object River is drawn (or, alternatively, the symbolic graphical object River must be in the background when the symbolic graphical object Forest is drawn). This example highlights how the visual representation maintains ambiguity problems, which are solved by adding information on the drawing process to the visual representation.

Two languages use the second approach (based on blackboard metaphor): Sketch [10] and Spatial-Query-By-Sketch [11]. In these languages the user draws a freehand visual representation of his query, as if on a blackboard, without explicit references to operators to be applied to geographical objects involved in the query. As with the other languages, in Sketch and Spatial-Query-By-Sketch too a query may have multiple interpretations. Spatial-Query-By-Sketch resolves the problem by considering and proposing to the user both the exact solution of the query and other approximate solutions obtained by removing or relaxing some relationships. In this manner Spatial-Query-By-Sketch includes multiple interpretations in the result, and it is the user's task to select the correct interpretation of his query. However, there is no guarantee that the different interpretations include the user's interpretation. For example, a topological relationship between two symbolic graphical objects can be removed by also considering other nearest topological relationships as acceptable. This approach allows, for example, to consider as an acceptable result a pair of touching geographical objects even if in the visual representation of the query they are disjoined.

In Spatial-Query-by-Sketch the query can be represented by one of the pictures of Figure 3-e considered as near to the desired result. In fact, each query produces all results that can be obtained by relaxing a limited number of relationships (River disjoint Forest, in this example). Obviously, each query produces a set of results containing the one actually required, which can be selected by the user. By selecting the result of interest the user specifies the correct interpretation of his query, removing all results with undesired relationships. Other types of ambiguities must be considered if a sketch represents the visual query. In fact, using this kind of approach, it is also necessary to recognize the different components that form the drawing. Such types of ambiguities are not discussed in this paper.

In [1] the authors confront the ambiguity problem in visual GIS query languages and propose a taxonomy based on user actions and system materialization (the different images the system can materialize), for distinguishing ambiguities, as well as ways of resolving them, and a model to solve a particular case of ambiguity. The proposed system, an enlargement of Lvis, establishes a dialog with the user; whenever an ambiguity occurs, it shows all the available configurations and requests a choice. As little research has yet been done in this area, ambiguities are still one of the most important and difficult problems in visual GIS query languages. The authors conclude that the strategy for avoiding ambiguities in most visual geographic query languages is to define not fully visual, but hybrid languages, including a textual part and offering a grammar with low expressive power.

Instead, in this paper a pictorial query language for geographical data is proposed fully pictorial without the necessity to establish a "dialog" between system and user, as well as the necessity of hybrid (textual-graphical) solutions.

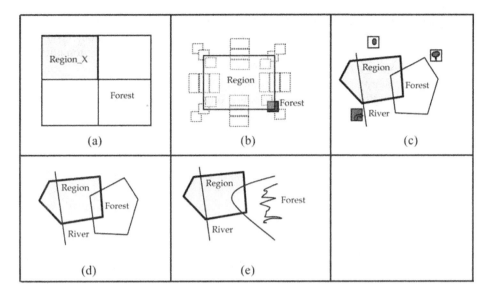

Fig. 3. The same visual query represented with different languages

4   The Geographical Pictorial Query Language (GeoPQL)

The Geographical Pictorial Query Language (GeoPQL, [12]) is an evolution of the Pictorial Query Language (PQL, [13]) and which resolves some PQL limitations by introducing two new operators (G-any and G-alias).

In reality G-any is not exactly an operator but rather a meta-operator, which translates itself in the pictorial query language into a set of all operators applicable to the previously drawn pair of sgo and to which G-any has been applied. For example, if it is drawn (applied) between the sgo River and Forest of figure 1-a, it produces all the valid topological relationships between an sgo polyline and an sgo polygon, i.e. the expressions "River *Disjoined From* Forest", "River *Touches* Forest" and "River *Passes-through* Forest". For simplicity, we will also apply the term "operator" to the meta-operator G-any.

In contrast, G-alias is the unique unary operator used by GeoPQL. It had to be introduced in order "to duplicate" an sgo so that queries in which the operator OR is present can be answered. For example, if the query is "Find all Regions which *include* a Lake OR which are *passed through* by a River", the user has to draw one polygon inside another polygon (lake and region respectively), and one disjoined polygon passed through by a polyline (Region and River respectively), to link between them, by the G-alias operator, the two Region polygons (thus declaring that the two polygons represent the same Region), and to define the query target (in our case, one Region polygon, even if the system will put in evidence both the Region polygons).

GeoPQL has various peculiarities. For example, it needs to explicitly express only three operators: G-any (to avoid expressing as many pictorial representations as there are possible applications of the valid operators to the pair of drawn sgo), G-distance

(distance) (to express the numeric value which specifies the constraint linked to the operator), and G-alias (to link two sgo representing the same duplicated objclass). All other operators are applied without being expressing, as they are automatically deduced by the query's pictorial representation.

Another peculiarity is that GeoPQL interfaces with ESRI's GIS ArcView® and improves and simplifies interaction by using browsing techniques that allow data actually present in the database to be selected.

The main results obtained in GeoPQL are: solution of ambiguities inherent to the spatial representation of a geographical query, exact matching between the query and the obtained results and high expressive power, because a user composes his query pictorially and univocally expresses in this manner all the topological relations among the symbolic graphical objects involved.

With GeoPQL, it is possible to specify queries using *symbolic graphical objects* (*sgo*) that have the appearance of the three classic types: point, polyline and polygon. The user can assign each sgo with a semantic linked to the different kinds of information (layer) in the geographical database. Specifically, constraints can be imposed on both the attributes of the geographical data and their topological position, and the query's target information (a specified layer or a set of layers) can be specified. In the following a layer is represented by a class and an element of a layer by a geographical object.

This fact allows to formulate queries on the geographical database simply by drawing a spatial representation of symbolic graphical objects, without knowing the database structure or the query language textual syntax.

Section 4.1 illustrates the basic concepts of the pictorial query language, and Section 4.2 proposes the set of operators used by GeoPQL.

4.1 Basic Concepts

An algebra consists of a set of formally defined operators and a set of data structures (or operands) which form its alphabet A. The operators, operating on one (unary) or two (binary) operands, produce results which can be either still elements of the alphabet A (closed algebra), or elements out side the alphabet (open algebra).

In the case of a pictorial query language the following are the alphabet's elements: point, polyline, polygon, oriented polyline, set of points, set of polylines, set of polygons, set of oriented polylines, empty. All elements which include the word "set" have a "cardinality"

In addition, the element "oriented polyline set" also has an "order" defined by the elements of this set.

In our case we consider only an alphabet subset A formed by the elements point, polyline, polygon and empty. The elements "set of points", "set of polylines" and "set of polygons" are considered equivalent to a unique sgo "point", or "polyline", or "polygon". We do not consider the oriented polyline (and the set of oriented polylines).

This means that the present version of GeoPQL is not able to satisfy constraints referring to the cardinality of a sgo (its cardinality is always 1) or the orientation of a polyline, and that some operators are used in to "select" symbolic graphical objects which satisfy the relationship of which the operator is an element, rather than to obtain the "result" of the operation.

This version thus considers the alphabet elements "point, polyline, polygon, empty" only, while considering the "set of the same type elements" as equivalent to only one element (of the same type of set). All this characterizes GeoPQL, remembering that the result of the application of one operator (for instance, *overlap*) to two operands (sgo, for instance, two *polygons* A and B)) is the polygon common to the two polygons A and B, but this result is used only as a condition (overlap empty-non empty) to select all the geographical objects which satisfy this condition (element not found – found). All the elements which satisfy this condition form the result of the pictorial query.

Note that, because the result of the operation is used as a selection condition for geographical objects in the geographical database, if the user stored this result, he would have to define a new class (layer) with a suitable name, and would also have to define if and which attributes are inherited from the original class (layer).

Below is a formal description of the data structure and set of operators used by GeoPQL.

An *sgo* is formally defined as a 4-tuple $\psi = \langle id, objclass, \Sigma, \Lambda \rangle$ where:

- *id* is the *sgo* identifier. The *id* is the code assigned in order to identify each sgo in queries in a project. A GeoPQL project is formed by a set of queries and the reference to the selected database for querying.
- *objclass* is the set (possibly empty) of class names iconized by ψ;
- Σ represents the attributes to which the user can assign a set of values; this allows selection to be made among the classes of objects or their instances, iconized by ψ. Some attributes can be referred to a temporal dimension. In particular, different kinds of temporal dimensions can be considered, represented by temporal intervals or instants. Σ is formed by the set of attributes of the classes represented by the *sgo*. The attributes of a class correspond to the attributes of the layer of the database represented by the class. In a query, the user can express the set of desired values for an attribute by an expression involving boolean and other operators, value, references to other attributes of the same and of other objects and classes.
- Λ is the ordered set of coordinate pairs (h, v), which defines the spatial extent and position of the *sgo* with respect to the coordinate reference system of the working area. Using Λ it is possible to determine spatial relationships among symbolic graphical objects.

From a geometric point of view, an *sgo* is defined by the boundary and the internal points of the extent of the sgo. The definition, evaluation and applicability of the operators are defined in terms of these characteristics.

Described below are the semantics and properties of the different operators, using the following symbolism: Let ψ be an *sgo,* then:

$\partial\psi$ = boundary of ψ is defined as:
- if ψ is a polygon, $\partial\psi$ is the set of its accumulation points;
- if ψ is a polyline, $\partial\psi$ is formed by its endpoints;
- if ψ is a point $\partial\psi$ is the empty set.

$\psi^o = \psi - \partial\psi =$ interior of ψ.

$Dim(\psi) = \quad 0 \quad$ if ψ is a point

$\qquad\qquad\quad 1 \quad$ if ψ is a polyline

$\qquad\qquad\quad 2 \quad$ if ψ is a polygon.

The *sgo* and the null element together form the elements of alphabet A. This alphabet, together with the set of operators defined below form the reference model.

4.2 The Set of Operators Used by GeoPQL

The set of operators used by GeoPQL (with the specification made at the beginning of Section 4) comprises of twelve operators: G-union, G-difference, G-disjunction, G-touching, G-inclusion, G-crossing, G-pass-through, G-overlapping, G-equality, G-distance, G-any and G-alias.

These operators, their semantics and the alphabet A, defined by the *sgo* <point, polyline, polygon, geo-null> where "geo-null" defines the empty object, form the system's geographical model. The set of operators was chosen in the light of the following goals: definition of a non-ambiguous language, exact matching between query and obtained results, high expressive power, interface with ArcView® (the most used geographical data management system). For this reason GeoPQL has greater expressive power and more operators than other visual languages for geographical data. The only exception is Spatial-Query-By-Sketch [11], which is characterized by a greater expressive power and number of operators, however this language does not ensure exact matching between the query and the obtained results, some results can be obtained by a set of different queries and conversely a query can obtain different results of which only one is the exact match. In addition, some results cannot be obtained by a query that is the exact match of the desired result, but are obtained only by approximate queries. Finally some operators such as G-alias and G-any are introduced only by GeoPQL.

In most queries on geographical data, expressions are required to express a set of topological relationships (with a rather obvious intuitive meaning). Below is a brief description of the expressions used in GeoPQL. All operators are applied to a pair of operands (sgo) (binary operations), except the G-alias which is applied to a single operator (unary operation). The sgo result of an expression, if not differently specified in the operator definition, is solely characterized by the point set defining the spatial extent and position.

G-union definition (Uni): Let ψ_i, $\psi_j \in$ A be two *sgo* of the same type (i.e., with the same dimension). The *G-union* of two *sgo* ψ_i and ψ_j produces the A element ψ_h, such that its point set is equal to the set union of the point sets of ψ_i and ψ_j.

Its semantics can be stated simply as follows:

Uni (ψ_i: A , ψ_j: A) $\rightarrow$ ψ_h: A

Semantics: Point(ψ_h) = Point(ψ_i) $\cup$ Point(ψ_j)

Where   dim(ψ_i) = dim (ψ_j).

G-touching definition (Tch): Let ψ_i, $\psi_j \in$ A be two *sgo* with dimensions not equal to 0 and all points common to the two ψ contained in the union of their boundaries.

The *G-touching* of two *sgo* ψ_i and ψ_j produces the A element ψ_h, such that its point set is equal to the set intersection of the point sets of the two sgo.

Its semantics can be stated simply as follows:

Tch (ψ_i: A , ψ_j: A) → ψ_h: A

Semantics: Point(ψ_h) = Point(ψ_i) ∩ Point(ψ_j)

Where: dim(ψ_i) = 1 or 2 and dim (ψ_j) = 1 or 2

Point(ψ_i) ∩ Point(ψ_j) ⊆ Point($\delta\psi_i$) ∪ Point($\delta\psi_j$)

G-inclusion definition (Inc): Let ψ_i, ψ_j ∈ A be two *sgo* with ψ_i with dimension not equal to 0 and greater than ψ_j and all points of ψ_j contained in ψ_i. The *G-inclusion* of two *sgo* ψ_i and ψ_j produces the A element ψ_h, such that its point set is equal to the point set of the sgo ψ_j.

Its semantics can be stated simply as follows:

Inc (ψ_i: A , ψ_j: A) → ψ_h: A

Semantics: Point(ψ_h) = Point(ψ_j)

Where: dim(ψ_i) = 1 or 2 and dim(ψ_i) >= dim(ψ_j)

Point(ψ_j) = Point(ψ_i) ∩ Point(ψ_j)

G-disjunction definition (Dsj): Let ψ_i, ψ_j ∈ A be two *sgo* of any type that do not have points in common. The G-disjunction of an sgo ψ_i from another sgo ψ_j produces the A element ψ_h=null.

Its semantics can be stated simply as follows:

Dsj (ψ_i: A , ψ_j: A) → ψ_h=null: A

Semantics: Point(ψ_h) = ∅

Where: Point(ψ_i) ∩ Point(ψ_j) = ∅

G-pass-through definition (Pth): Let ψ_i, ψ_j ∈ A be two *sgo* with ψ_j having dimension equal to 2 and ψ_i having dimension equal to 1 and internal points in common to both the boundary and internal points of ψ_j. The G-pass-through of an sgo ψ_i with respect to another sgo ψ_j produces the A element ψ_h with dim(ψ_h) = 1, such that its point set is equal to the set intersection of the point sets of the ψ_i and ψ_j sgo.

Its semantics can be stated simply as follows:

Pth (ψ_i: A , ψ_j: A) → ψ_h: A

Semantics: Point(ψ_h) = Point(ψ_i) ∩ Point(ψ_j)

Where: dim(ψ_i) = 1 and dim(ψ_j) = 2

(Point ($\psi^o{}_i$) ∩ Point ($\partial\psi_j$) ≠ ∅) ∧ (Point ($\psi^o{}_i$) ∩ Point ($\psi^o{}_j$) ≠ ∅)

G-difference definition (Dif): Let ψ_i, ψ_j ∈ A be two sgo of the same type. The G-difference between an sgo ψ_i and another sgo ψ_j produces the A element ψ_h such that its point set is equal to the point set of ψ_i which are not points of ψ_j.

Its semantics can be stated simply as follows:

$Dif(\psi_i: A, \psi_j: A) \rightarrow \psi_h: A$

Semantics: Point(ψ_h) = Point(ψ_i) - Point(ψ_j)

G-crossing definition (Crs): Let ψ_i, $\psi_j \in A$ be two sgo having dimension equal to 1 and single internal points in common. The G-crossing between an sgo ψ_i and another sgo ψ_j produces the A element ψ_h with dim(ψ_h) = 0, such that its point set is equal to the set of points of $°\psi_i$ which are in common to $°\psi_j$.

Its semantics can be stated simply as follows:

$Crs(\psi_i: A, \psi_j: A) \rightarrow \psi_h: A$

Semantics: Point(ψ_h) = Point($\psi^\circ i$) $\cap$ Point($\psi^\circ j$)

Where: dim(ψ_i) = 1 and dim(ψ_j) = 1

Point($\psi^\circ i$) $\cap$ Point($\psi^\circ j$) $\neq \varnothing$ and dim (ψ_h) = 0

G-overlapping Definition (Ovl): Let ψ_i, $\psi_j \in A$ be two *sgo* of the same type, with internal points in common. The G-overlapping between an sgo ψ_i and another sgo ψ_j produces the A element ψ_h with the same dimension, such that its point set is equal to the set of points of ψ_i which are in common to ψ_j.

Its semantics can be stated simply as follows:

$Crs(\psi_i: A, \psi_j: A) \rightarrow \psi_h: A$

Semantics: Point(ψ_h) = Point(ψ_i) $\cap$ Point(ψ_j)

Where: dim(ψ_i) > 0, and dim(ψ_j) > 0 and dim(ψ_i) = dim(ψ_j)

Point($\psi^\circ i$) $\cap$ Point($\psi^\circ j$) $\neq \varnothing$

G-equality definition (Eql): Let ψ_i, $\psi_j \in A$ be two *sgo* of the same type, with their boundary and internal coincident. The G-equality between an sgo ψ_i and another sgo ψ_j, with dim(ψ_i)=dim (ψ_j), produces the A element ψ_h with the same dimension such that its point set is equal to the point set of ψ_i and ψ_j.

Its semantics can be stated simply as follows:

$Eql(\psi_i: A, \psi_j: A) \rightarrow \psi_h: A$

Semantics: Point(ψ_h) = Point(ψ_i) = Point(ψ_j)

Where: dim(ψ_i) = dim(ψ_j)

Point($\psi^\circ i$) = Point($\psi^\circ j$) $\neq \varnothing$

Point($\partial\psi i$) = Point($\partial\psi j$) $\neq \varnothing$

The various distances of the distance operator, e.g. minimum, maximum, may be considered, each with a different calculus function. This operator can be used to find all *sgo* having distance θ (θ being one of the following symbols: >, <, =, ≤, ≥, ≠) from the reference *sgo*.

G-distance definition (Dst): Let ψ_i, $\psi_j \in A$ be two *sgo* of any type that do not have points in common. The G-distance of an sgo ψ_i respect to another sgo ψ_j produces the A element ψ_h such that its point set is equal to the point set of ψ_i, and the set of attribute of ψ_h is set of attributes of ψ_i added to the Distance Measure$_\phi$ attribute. In the following ϕ is a qualifier, which specifies the kind of distance (minimum, maximum, etc). The present version of the prototype considers the minimum distance only.

Its semantics can be stated simply as follows:

$Dst_\phi\,(\psi_i\colon\ \mathrm{A}\,,\ \psi_j\colon\ \mathrm{A}\,) \rightarrow \psi_h\colon\ \mathrm{A}$

Semantics: $\mathrm{Point}(\psi_h) = \mathrm{Point}(\psi_i)$

$\mathrm{Attr}(\psi_h) = \mathrm{Attr}(\psi_i) + \text{Distance Measure}_\phi$

Where: $\mathrm{Point}(\psi_i) \cap \mathrm{Point}(\psi_j) = \varnothing$

The new attribute representing the distance specification allows to specify a selection expression that includes conventional operators ($>$, $<$, $=$, $\neq$, etc.) or methods that behave like operators. GeoPQL uses the value of this measure to verify the constraint expressed in the query in order to select (or not) the object that satisfies this constraint respect to the reference object. For example, the query "Select all the English cities which have a distance > 100 miles from London" is an example of this situation.

The G-distance operator is graphically represented in the pictorial query by the following symbol $\leftarrow \rightarrow$.

G-alias Definition (Als): Let ψ_i be an sgo of any type. ψ_j is an *alias* of ψ_i if it is the same sgo with a spatial translation (and consequently in relationship with other *sgo*).

Its semantics can be stated simply as follows:

$Als\,(\psi_i\colon\ \mathrm{A}) \rightarrow \psi_h\colon\ \mathrm{A}$

Semantics: $\mathrm{Point}(\psi_h) = \mathrm{Point}(\psi_i)$

$\mathrm{Attr}(\psi_h) = \mathrm{Attr}(\psi_i)$

where $\dim(\psi_h) = \dim(\psi_i)$

The G-alias operator is graphically represented in the pictorial query by means of the following symbol $\leftarrow \text{ALIAS} \rightarrow$. G-alias gives a second (or successive) representation of the same *sgo* to allow the possibility of expressing alternative sets of relationships. In practice, G-alias allows implementation of a query with the OR operator between two expressions in which the same *sgo* is used.

For example, in Figure 4 there is a pictorial representation in which regions can be *passed through* by a river or (alternatively) they can *include* a lake.

G-any Definition (Any): Let ψ_i, $\psi_j \in \mathrm{A}$ be two sgo of any type. The G-any between an sgo ψ_i and another sgo ψ_j produces a set of A element ψ_{h_n} (with $n=1, 2, 3$), each having elements of the same type as ψ_i and which are the result of all the valid expressions from the valid relationships between the two sgo ψ_i and ψ_j.

Its semantics can be stated simply as follows

$Any\,(\psi_i\colon\ \mathrm{A}\,,\ \psi_j\colon\ \mathrm{A}\,) \rightarrow \{\psi_{h_n}\}\colon\ \mathrm{A}$

Semantics: $\{\psi_{h_n}\}\colon\ \{\psi_{h_0}\} \cup \{\psi_{h_1}\} \cup \{\psi_{h_2}\}$

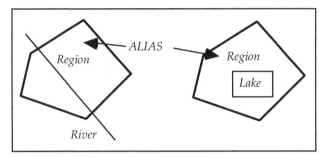

Fig. 4. A pictorial representation of a query with the G-alias operator

where   n = 1, 2, 3 and the elements of ψh_n of the same type being n fixed.

$$\{\psi h_0\} = \{\psi h_{0,1}, \psi h_{0,2}, \ldots, \psi h_{0,k}\}$$

In practical the application of the G-any operator allows to consider the different drawings of valid configurations between a pair of sgo, as in Figure 2 for River and Forest, all together in a query. Each drawing can produce results of different dimension. If the G-any operator is applied between two sgo any admissible relationship is valid between them.

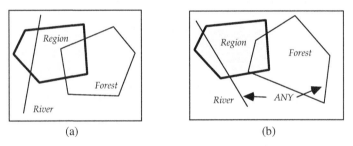

         (a)   (b)

Fig. 5. A pictorial representation of a query with the G-any operator

The G-any operator is graphically represented in the pictorial query by the symbol ←ANY→. To explain the use of the G-any operator, consider the following query: "Find all the regions which are *passed through* by a river and *overlap* a forest", where the user has no interest in the relationship between river and forest. If the user draws the configuration of Figure 5-a in such a way that the forest and the river are "Disjoined", then the system must interpret the query as "Find all the regions which are *passed through* by a river and *overlap* a forest, *and the forest is disjoined from the river*". Other query representations imply a spatial relationship between the forest and the river.

By the introduction of the G-any operator, the query is correctly drawn as in Figure 5-b, and it is possible to give the exact interpretation of the query with respect to the answer desired by the user. The G-Any operator allows visual representation of the need to remove relationships (as addressed in Section 2) between a pair of symbolic graphical objects and corresponds to set ψ_i in background state when ψ_j is drawn (supposing that ψ_j is drawn after ψ_i) as in the approach presented in Section 3.

5   Syntactic and Semantic Correctness

One important issue is the definition of a sound method for analysis of syntactic and semantic correctness of queries, which may lead to multiple system and user interpretations. For this reason, the paper illustrates an approach able to determine the exact syntactic and semantic interpretations of geographic configurations involved in queries expressed by a pictorial query language.

The paper will focus on the syntactic and semantic correctness of spatial configurations in the context of the declarative geographic pictorial query language, GeoPQL [12][13].

A visual query language with clear syntax and semantics can prevent a priori many ambiguities, minimizing multiple interpretations. The goal of this paper is to present the configurations between symbolic graphical objects which can be considered syntactically correct in a geographical context, identify the set of GeoPQL operators referred to each configuration (Section 5.1), and give each configuration a non-ambiguous semantic (Section 5.2).

5.1   The Syntactic Correctness of Pictorial Configurations

In this section a generic sgo pair, part of the set of all query sgo, is considered. The set of operators syntactically admissible for this pair is defined, starting from possible spatial configurations (Figure 6). In this way the system considers for each configuration drawn by the user only the syntactically correct expressions, ensuring the syntactical correctness of the pictorial query. The proposed operators represent the relationships on the spatial properties of the objects (or classes) of the database that the user must specify in his query in order to find the geographic objects of interest.

Each possible spatial configuration is given a code of three alphabetic characters followed by a number. The first two characters indicate the type of sgo, the third indicates the type of spatial configuration, and the number distinguishes the configuration. In this manner the first configuration between two regions of Figure 6 is referred to as "aaA1", the last as "aaE1".

Let ψ_i and ψ_j be two symbolic graphical objects, which form a given configuration and are associated respectively with the database's geographical classes gc_α and gc_β, possibly coincident. For each configuration of Figure 5, a set of *syntactically correct* expressions <ψ_i *Operator* ψ_j> on the basis of geometric and topological properties of ψ_i and ψ_j can be considered.

Table 1 summarizes syntactically correct expressions for all the configurations of Figure 6.

The operators G-alias, G-any and G-distance, in contrast to the other operators, must be expressed in the query using a suitable symbol (a labeled edge).

Consequently, these operators are not linked to the particular configuration between the sgo pair involved, but to the symbols between them. For this reason, these kinds of operators require verification of the applicability requisites only, without considering the syntactical correctness of the configurations.

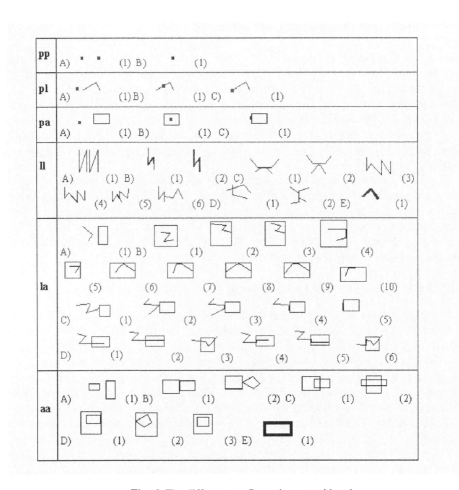

Fig. 6. The different configurations considered

5.2 The Semantic Correctness of Pictorial Configurations

In the previous section, all possible configurations of pairs of symbolic graphical objects (operands) were considered. Their syntactic correctness was verified and the set of applicable operators identified. These properties are independent of the geographic database. In contrast with syntactic correctness, semantic correctness is related to the meaning of the geographical information, so some syntactically correct

Table 1. Summary of syntactically correct expressions for configurations of Figure 4

Configurations	Descriptions	Operators syntactically correct
ppA1	The configuration represents two points ψ_i and ψ_j which are separate, or specifies a property of their union.	ψ_i Uni ψ_j (or ψ_j Uni ψ_i) ψ_i Dsj ψ_j (or ψ_j Dsj ψ_i)
ppB1	The points ψ_i and ψ_j are elements of different thematisms (classes) and their common property is their identical spatial coordinates	ψ_i Eql ψ_j (or ψ_j Eql ψ_i)
plA1	The configuration represents a point ψ_i and a polyline ψ_j which are separate.	ψ_i Dsj ψ_j (or ψ_j Dsj ψ_i)
plB1	In this configuration a point ψ_i is located over a polyline and ψ_j.	ψ_j Tch ψ_i ψ_j Inc ψ_i
plC1	This is equal to the previous configuration and represents a point ψ_i located over a polyline ψ_j on the boundary.	ψ_j Tch ψ_i ψ_j Inc ψ_i
paA1	The configuration represents two polygons ψ_i and ψ_j which are separate.	ψ_i Dsj ψ_j (or ψ_j Dsj ψ_i)
paB1	The configuration represents a point ψ_i located in a polygon ψ_j.	ψ_j Inc ψ_i
paC1	The configuration represents a point ψ_i located on the boundary of the polygon ψ_j.	ψ_j Tch ψ_i ψ_j Inc ψ_i
llA1	The configuration represents two polylines ψ_i and ψ_j which are separate, or allow specification of a property of their union.	ψ_i Dsj ψ_j (or ψ_j Dsj ψ_i) ψ_i Uni ψ_j (or ψ_j Uni ψ_i)
llB1-llB2	Two configurations can be considered: 1) ψ_j is completely within ψ_i ; 2) ψ_j is completely within ψ_i and they have one common boundary point.	ψ_i Uni ψ_j (or ψ_j Uni ψ_i) ψ_j Inc ψ_i ψ_j Dif ψ_i ψ_i Ovl ψ_j (or ψ_j Ovl ψ_i)
llC1- llC2- llC3- llC4- llC5- llC6	These six configurations represent the different cases in which the two polylines ψ_i and ψ_j can have some but not all of their points in common, without crossing.	ψ_i Tch ψ_j (or ψ_j Tch ψ_i) for configurations llC2, llC3 and llC6 ψ_i Uni ψ_j (or ψ_j Uni ψ_i) for all configurations ψ_j Dif ψ_i for configurations llC1, llC4, and llC5 ψ_i Ovl ψ_j (or ψ_j Ovl ψ_i) for configurations llC1, llC4, and llC5
llD1	These two configurations represent the different cases in which the two polylines ψ_i and ψ_j can have some but not all of their points in common, while crossing.	ψ_i Uni ψ_j (or ψ_j Uni ψ_i) for all configurations ψ_i Ovl ψ_j (or ψ_j Ovl ψ_i) for the configuration llD2 ψ_i Crs ψ_j (or ψ_j Crs ψ_i) for the configuration llD1

Table 1. (*continued*)

Configura-tions	Descriptions	Operators syntactically correct
llE1	In this configuration the polylines ψ_i and ψ_j are elements of different thematisms (classes) and are spatially coincident.	ψ_i Eql ψ_j (or ψ_j Eql ψ_i)
laA1	The configuration represents a polyline ψ_i and a polygon ψ_j which are separate.	ψ_i Dsj ψ_j (or ψ_j Dsj ψ_i)
laB1- laB2- laB3- laB4- laB5- laB6- laB7- laB8- laB9- laB10	These ten configurations represent the various cases in which a polygon ψ_j contains a polyline ψ_i.	ψ_j Inc ψ_i for all configurations ψ_i Tch ψ_j (or ψ_j Tch ψ_i) for configurations laB2 laB3-laB4-laB5-laB6-laB7-laB8-laB9-laB10
laC1- laC2- laC3- laC4- laC5	These five configurations represent the various cases in which a polyline ψ_i touches a polygon ψ_j.	ψ_i Tch ψ_j (or ψ_j Tch ψ_i) for all configurations ψ_i Dif ψ_j for configurations laC3 and laC4 ψ_j Inc ψ_i for the configuration laC5
laD1- laD2- laD3- laD4- laD5- laD6	These six configurations represent the cases in which a polyline ψ_i intersects a polygon ψ_j.	ψ_i Pst ψ_j for all configurations ψ_i Dif ψ_j for all configurations ψ_i Tch ψ_j (or ψ_j Tch ψ_i) for configurations laD4-laD5-laD6
aaA1	The configuration represents two polygons ψ_i and ψ_j which are separate, or allow a property of their union to be specified.	ψ_i Dsj ψ_j (or ψ_j Dsj ψ_i) ψ_i Uni ψ_j (or ψ_j Uni ψ_i)
aaB1- aaB2	These configurations represent the cases in which two polygons ψ_i and ψ_j are touching.	ψ_i Uni ψ_j (or ψ_j Uni ψ_i) ψ_i Tch ψ_j (or ψ_j Tch ψ_i)
aaC1- aaC2	These configurations represent the cases in which two polygons ψ_i and ψ_j overlap.	ψ_i Uni ψ_j (or ψ_j Uni ψ_i) ψ_i Dif ψ_j (or ψ_j Dif ψ_i) ψ_i Ovl ψ_j (or ψ_j Ovl ψ_i)
aaD1- aaD2- aaD3	These configurations represent the cases in which the polygon ψ_i encloses the polygon ψ_j.	ψ_i Uni ψ_j (or ψ_j Uni ψ_i) for all configurations ψ_i Dif ψ_j for all configurations ψ_i Inc ψ_j for all configurations ψ_i Ovl ψ_j (or ψ_j Ovl ψ_i) for all configurations ψ_i Tch ψ_j (or ψ_j Tch ψ_i) for aaD1 and aaD2
aaE1	The polygons ψ_i and ψ_j are elements of different classes and are spatially coincident.	ψ_i Eql ψ_j (or ψ_j Eql ψ_i)

configurations could be semantically incorrect, while some operators syntactically applicable to a configuration may be inapplicable semantically, due to the geographical information involved.

For example, the crossing of two polylines is always syntactically correct, but is semantically correct if the polylines represent two streets or a street and a river (the result could be a bridge), but is semantically incorrect if they represent two rivers. It is obvious that semantic correctness is a subset of the syntactic correctness of the GeoPQL operators applied to all possible pairs of symbolic graphical objects.

Let ψ_i and ψ_j be two symbolic graphical objects which form a given configuration and are associated respectively with the database's geographical classes gc_α and gc_β, possibly coincident.

$<\psi_i\ Operator\ \psi_j>$ is *semantically correct* with regard to the selected geographical database if it is syntactically correct (for the configuration) and the result in a non-null set of geographical objects of the selected database. A configuration between the symbolic graphical objects ψ_i and ψ_j is *semantically correct* if it is syntactically correct and at least one of its associated operators is semantically correct. Obviously, semantic correctness depends on the geographical classes gc_α and gc_β associated with ψ_i and ψ_j.

For all operators for which the symmetric property is not syntactically valid, semantic correctness depends on the order of the two operands. For example, a region can include a lake, but the converse is false.

If the result is a null set of objects, however, the user can autonomously define $<\psi_i\ Operator\ \psi_j>$ as *"semantically correct, but absent in the database"* for that configuration and manage "lists of semantic correctness for configurations, operators and pairs of geographical classes" or he can define $<\psi_i\ Operator\ \psi_j>$ as *"semantically incorrect"* and manage "lists of semantic incorrectness for configurations, operators and pairs of geographical classes".

This procedure can be applied every time a new geographic class is defined. However, it can be very onerous because it is necessary to specify for all configurations and operators between the new class and previously defined classes giving a null result, whether they are correct but absent or incorrect.

For this reason, the system considers configurations and operators giving a non-null result as semantically correct. It considers all remaining configurations as "undetermined" until the user formulates a query which involves a $<\psi_i\ Operator\ \psi_j>$ expression for the pair of geographical classes. It is only then that the user decides if this construct is semantically correct but absent or incorrect.

It is also important to specify the different semantic role played by the topological operators with respect to the other operators. For example, if the query consists of two polylines, which cross each other, the syntactically correct expressions for this configuration are:

ψ_i Crs ψ_j

ψ_i Uni ψ_j

The first expression derives from a topological property of the configuration of the two symbolic graphical objects involved in the query. For this reason, it is semantically correct if the user explicitly specifies that the operator is semantically correct for that pair of operands, or if there is a set of geographical object pairs which satisfies this expression in the database.

The second expression is taken in consideration only if the user specifies in it properties concerning the union of the two symbolic graphical objects. For this reason it is semantically correct if there are common properties between the two geographical classes gc_α and gc_β represented by ψ_i and ψ_j.

6 Resolution of Ambiguities

In a visual query representation, as shown in Figure 1, it is impossible not to explicitly represent at least one relationship between a pair of sgo, independently of whether a relationship must be represented or not in the query. For this reason, to obtain an unambiguous visual query language it must be possible to eliminate such relationships from the query's visual representation.

This is permitted by G-any. If such a relationship is defined between a pair of sgo, it means that no relationship exists between them. For example, consider once more the query: "Find all the regions which are *passed-through* by a river and *overlap* a forest", where the user has no interest in the relationship between river and forest. G-any resolves some ambiguities relating to the query formulation through different representations (see Figure 1), each having different relationships between forest and river.

Each query in Figure 1 can also be interpreted in different ways. The correct interpretation of Figure 1-a is "Find all the regions which are *passed-through* by a river and *overlap* a forest *and in which the river is disjoined from the forest*". The correct interpretation of the query of Figure 1-b is "Find all the regions which are *passed-through* by a river and *overlap* a forest *and in which the river touches the forest*", while Figure 1-c is correctly interpreted as "Find all the regions which are *passed-through* by a river and *overlap* a forest *and in which the river passes-through the forest*". By introducing the G-any operator, the query is correctly drawn as shown in Figure 7-a and all relationships are correct, whether the river passes through the forest, is adjacent to the forest, or is disjoined from the forest.

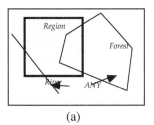 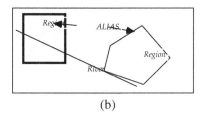

| (a) | (b) |

Fig. 7. The use of the G-any and G-alias operators

Note that the definition of the operator G-any also allows visually very complicated queries to be expressed. Suppose that queries are formulated with more than three objects (six, seven, etc.). The user may be interested in just a limited set of relationships between pairs of symbolic graphical objects, and for other pairs any relationship is good. With the G-any operator, the user is able to define the query exactly with respect to what he wishes to obtain as answer.

However, it is not always true that only one relationship exists between a pair of objects or that all of them are true. In many cases it must be possible to express one OR another given set of relationships. For this reason the G-alias operator is proposed. This allows a query to be implemented with the OR operator (in practice G-alias duplicates an *sgo* in order to draw a query in which it is used in the two alternative parts). G-alias also resolves some ambiguities, which may arise in interpreting the query's pictorial configuration. For example, Figure 7-b shows the pictorial representation of the query: "Find the regions which are *passed through* OR are *touched* by a river". Without the G-alias operator the query is ambiguous and may give rise to different interpretations.

Another problem in query interpretation derives from the possible need to consider the symbolic graphical objects obtained by applying operators to pairs of sgo and their mutual relationships, and then applying operators to the obtained symbolic graphical objects. The user does not want most of these relationships. For this reason the only ones considered by the system are those derived from the drawing. Nevertheless, it is possible to transform the result of an operator to a pair of sgo (or a sequence of operators) in a new virtual sgo and evaluate its relationships with other sgo during the query elaboration. A virtual sgo is not directly drawn by the user but obtained as a result of an operator to a pair of sgo (or a sequence of operators) relationships with one or more sgo need to be considered for it. However these multi-level relationships are considered only if explicitly required by specifying pairs of virtual sgo and sgo. By default the only relationships considered are those between pairs of drawn sgo.

Beside the classical situation previously illustrated in Figure 1 (and also discussed in [1]), other ambiguous situations can occur. For example, always with regard to two polygons and one polyline, the relationships existing between a symbolic graphical object obtained as result of a previous operation between two symbolic graphical objects drawn by the user and another (third) symbolic graphical object of the query. Suppose, then, to have two polygons A and B which are in overlapping between them, and suppose that a polyline P *Passed-through (PTH)* the two polygons. Several different situations between the polyline and the intersection of A and B can be considered, shown in Figure 8.

In such configurations the relationship between the polyline P and the result of A OVL B (called X), can be:

P DSJ X (Figure 8-a)
P TCH X (Figure 8-b)
P PTH X (Figure 8-c)

The overlap of A and B could be considered as a virtual symbolic graphical object, in fact it is not an object directly drawn by the user but obtained by applying an operator to a pair of objects. This means that they could be considered as "second level", "third level", … "n-th level" relationships which involve virtual symbolic graphical objects that represent the intermediate operations (in the previous case, A OVL B).

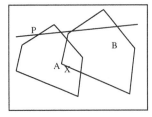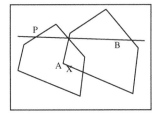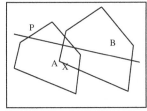

Fig. 8. Representation of relationships between the polyline P and the intersection of A and B

The number of levels depends on the number of symbolic graphical objects that form the query, as well as on the topological relationships that exist between such objects. It could be very complicated to consider all these relationships in the query and, in any case the user is usually interested in just a small subset of them.

For this reason, if not explicitly specified by the user, the relationships involving virtual symbolic graphical objects are not considered. So in the previous example in which the polyline passes-through the polygons A and B, the configurations obtained are those verifying the relationships:

A OVL B
Polyline PTH A
Polyline PTH B

Independently of the relationship between the polyline and the virtual symbolic graphical object X.

GeoPQL manages the problem of expressing queries involving just some virtual symbolic graphical objects by permitting specification of the smallest parts of a symbolic graphical object or a combination of them, which can be obtained using an expression involving symbolic graphical objects and operators.

So for example the query in Figure 9-a has just three symbolic graphical objects A, B, and C, but the virtual symbolic graphical objects 1, 2, 3, 4, 5, 6, and 7 or any combination of them could also be considered, such as the combination of 6 and 7, corresponding to A OVL B. The user must specify if and which of these virtual symbolic graphical objects (and their relationships) should be considered. This operation is performed by selecting one or more of the smallest symbolic graphical objects and specifying that the obtained object is a virtual symbolic graphical object of the query.

Analogously for polylines, in Figure 9-b GeoPQL considers the objects A, B, and C, as well as the sub-objects 1, 2, 3, x, y, and z.

GeoPQL does not currently distinguish between geographical objects formed by one or more than one polygon (i.e. set of polygons), polyline (i.e. set of polylines) and point (i.e. set of points). Neither can the cardinality of set of polygons, set of polylines, and set of points be considered. For this reason, the symbolic graphical object polygon (or polyline, or point) represents both: i) geographical objects described by one polygon (or polyline, or point) only and ii) geographical objects described by a set of two or more polygons (or polylines, or points). Similarly GeoPQL does not consider the cardinality of the sgo obtained as result of an operator. In any case, the only limitation concerns the possibility of distinguishing the

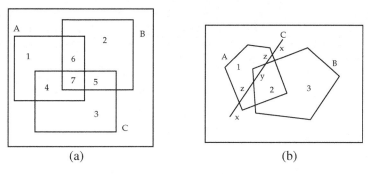

(a)   (b)

Fig. 9. Representation of queries and symbolic graphical objects in their entirety and in their smallest parts

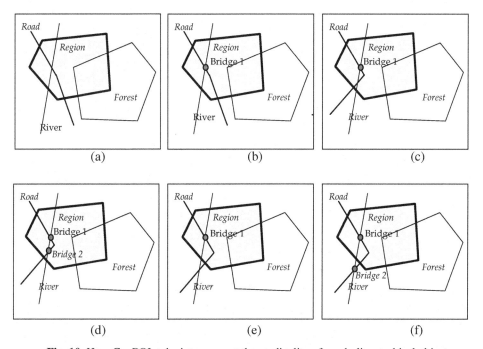

(a)   (b)   (c)

(d)   (e)   (f)

Fig. 10. How GeoPQL take into account the cardinality of symbolic graphical objects

cardinality of an sgo. So for example, the query in Figure 10-a is "Find all regions which are *passed through* by a river and a road and *overlap* a forest, where the road *crosses* the river and both road and river are *disjointed* from the forest". Considering Figures 10-b, 10-c, 10-d and 10-e, the four queries are equivalent for GeoPQL and mean: "Find all regions which are *passed through* by a river and a road and *overlap* a forest, where the road *crosses* the river in the region but not in the forest and the river and the road are *disjointed* from the forest". In this query the object Bridge 1 (and Bridge 2 in figure 10-d) obtained applying the G-cross operator has been highlighted as a new object of the query, and so it is necessary to consider its relationship with the other objects. In contrast, the query in Figure 10-f is not equivalent to the previous queries, because the two crossings between the road and the river have different spatial relationships. In fact the Region includes Bridge 1 but not includes Bridge 2. This query means: "Find all regions which are *passed through* by a river and a road and *overlap* a forest, where the road *crosses* the river both in and out of the region but not in the forest and the river and the road are *disjointed* from the forest".

7   Pictorial Queries by GeoPQL

When the target is defined, the execution of the query, requested by the user, is carried out by the system and the result, both as map and as object list, is given after just a few seconds.

The query described by a set of spatial relationships is translated and visualized to the user in an eXtended SQL language, called XSQL. The textual query is continuously updated during the drawing phase, and it follows modifications, deletions and shifting of the pictorial query. The query is translated into ArcView® and executed on ArcMap® (the geographical database of ArcView®). The user thus obtains a layer on which the geographical objects that satisfy the query are stored and visualized.

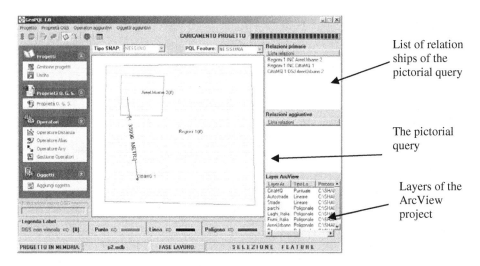

List of relation ships of the pictorial query

The pictorial query

Layers of the ArcView project

Fig. 11. A Query Example step-by-step: the pictorial query

All ArcView®'s basic browsing and drawing functions are present in the implemented system, as well as the typical functions defined in GeoPQL, very powerful and transparent geo-processing and visual analysis of the drawing in real time. The query is processed through visual selection of a target and results can be

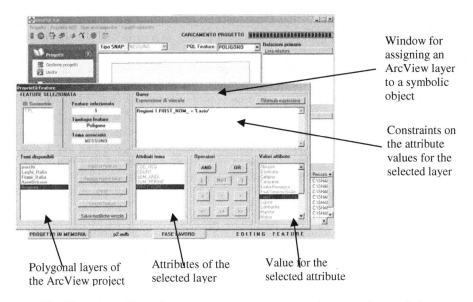

Window for
assigning an
ArcView layer
to a symbolic
object

Constraints on
the attribute
values for the
selected layer

Polygonal layers of
the ArcView project

Attributes of the
selected layer

Value for the
selected attribute

Fig. 12. A Query Example step-by-step: assigning constraints to attributes of a layer

pre-visualised through "results preview", enabling their development to be followed. Figure 11 shows the main window of GeoPQL, which has an area for pictorial query formulation, an area in which the user can browse the layers of the loaded ArcView® Project and an area in which the user can browse the relationships of the pictorial query.

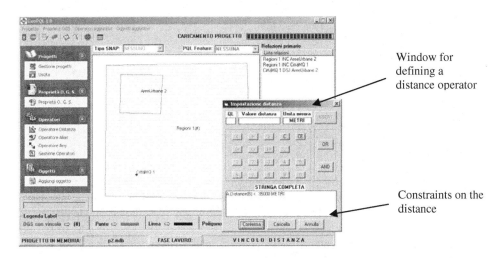

Window for
defining a
distance operator

Constraints on the
distance

Fig. 13. A Query Example step-by-step: defining a distance operator

The query formulated in the pictorial area is "Select all cities of the Lazio region disjointed from the metropolitan area of Rome and less than 35 Km from the metropolitan area of Rome".

Through the window of Figure 12 the user can assign an ArcView® layer to a symbolic graphical object and specify constraints concerning the attribute values for the symbolic graphical object. This operation is performed according to the layers of the ArcView® project browsing possible layers for the kind of symbolic graphical object (a polygon), attributes for the selected layer and values for the selected attribute. The user can also define a distance operator (Figure 13) specifying constraints by a formula. Figure 14 shows the result window of GeoPQL. This window visualizes a preview of the result of the pictorial query formulated (iconized) and its translation in XSQL.

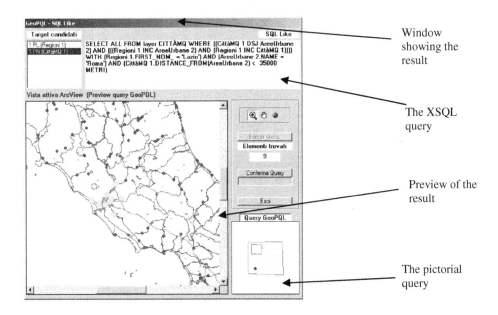

Fig. 14. A Query Example step-by-step: visualization of the result

Figure 15-a shows a query with a G-Alias operator. The query formulated is "Select all metropolitan areas belonging to a region and crossed by or touching a river". The two blocks of relationships are in OR between them. The result of the previous query is shown in Figure 15-b. Note that rivers are represented in database as polygons.

Figure 16-a shows a query with G-Alias and G-Any operators. The query formulated is "Select all regions overlapping a lake or that do not have highways". The result of the previous query is shown in Figure 16-b.

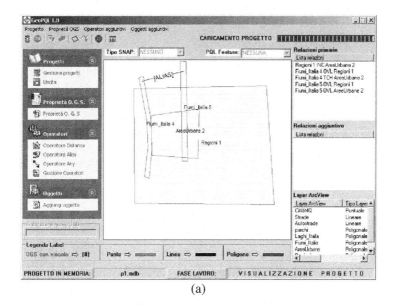

(a)

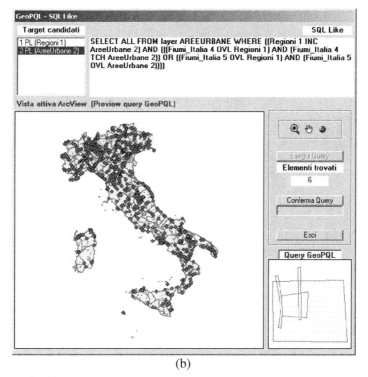

(b)

Fig. 15. A Query Example with a G-alias operator (a) and the result visualization (b)

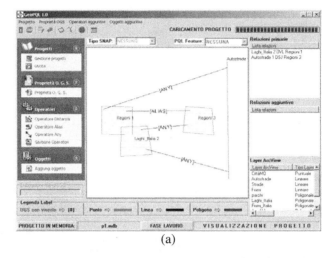

(a)

(b)

Fig. 16. A Query Example with G-alias and G-any operators (a) and the result visualization (b)

8 Conclusions

This paper presented the Pictorial Geographical Query Language GeoPQL, which enables easy query formulation. It discussed some ambiguities which can happen in a pictorial query interpretation and in which way GeoPQL resolves such ambiguities. In particular, the paper proposed two operators, G-any and G-alias, able to help the user to have a unique answer, giving a correct interpretation of his query.

The main results obtained in GeoPQL are:

• A high expressive power, because a user composes pictorially his query and thus expresses all topological relations among the symbolic graphical objects involved. A user focuses his attention on symbolic graphical objects and on their topological relationships, whatever geographic class or object they represent. The non-procedural characteristics of this language allow to devote the user attention to what he wants and not how to obtain it.

• The solution of ambiguities inherent to the spatial representation of a geographical query. In fact, the query is interpreted considering all relationships between symbolic graphical objects of the picture, however differently from other languages it is possible to remove or modify some undesired relationships, using the operators introduced in GeoPQL.

• The exact match between the query and obtained results. In GeoPQL the results obtained are only those satisfying the relationships, with irrelevant relationships removed from the picture directly by the user.

The authors are now focusing research activity in two areas: to enlarge the set of operators including directional, temporal and other measurement operators and to verify more interactive approaches (sketch, speech, …) in line with the growth in multimodal interaction tools.

References

1. Favetta F., Aufaure-Portier M.A. "About ambiguities in visual GIS query languages: a taxonomy and solutions" VISUAL 2000, LNCS n. 1929, Springer-Verlag Publ., pp.154-165, 2000
2. Papadias D., Sellis T. "A Pictorial Query-by-Example Language" Journal of Visual Languages and Computing, Vol.6, N.1, pp. 53-72, 1995.
3. Kaushik S., Rundensteiner E.A. "SVIQUEL: A Spatial Visual Query and Exploration Language" 9th Intern. Conf. on Database and Expert Systems Applications - DEXA'98, LNCS N. 1460, pp. 290-299, 1998.
4. Mainguenaud M. "GROG: Geographical queries using graphs" Advanced Database Systems Symposyum, Japan Society, Kyoto - Japan, Dec. 1989
5. Mainguenaud M., Portier M.A. "Definition og Cigales: a GIS Query Language" Int. Conf. DEXA '90, Springer Verlag Publ., pp.275-280, 1990.
6. Calcinelli D., Mainguenaud M. "Cigales, a visual language for geographic information system: the user interface" Journal of Visual Languages and Computing, Vol. 5, N. 2, pp.113-132, 1994.
7. Aufaures-Portier M.A., Bonhomme C. "A High Level Language for Spatial Data Management" Third Int. Conf. on Visual Information Systems - VISUAL '99, LNCS N.1614, pp.325-332, 1999.
8. Calcinelli D., Mainguenaud M. "The Management of the ambiguities in a graphical query language for geographical information systems" Advances in Spatial Databases, 1991, LNCS n. 525, Springer-Verlag Publ., pp. 99-118
9. Lee Y. C., Chin F. L.: "An Iconic Query Language for Topological Relationship in GIS" International Journal on Geographical Information Systems, Vol. 9, n.1, pp 25--46, 1995
10. Meyer B. "Beyond Icons: Towards New Metaphors for Visual Query Languages for Spatial Information Systems" First Intern. Workshop on Interfaces in Database Systems, Springer-Verlag Publ., pp.113-135, 1993

11. Egenhofer M.J. "Query Processing in Spatial-Query-by-Sketch" Journal of Visual Languages and Computing, Vol. 8, N. 4, pp. 403-424, 1997
12. Ferri F., Rafanelli M. "GeoPQL: a Geographical Pictorial Query Language" Tech. Rep. IASI-CNR, 2004.
13. Ferri F., Massari F., Rafanelli M. "A Pictorial Query Language for Geographic Features in an Object-Oriented Environment". Journal of Visual Languages and Computing, Vol. 10, N. 6, pp. 641-671, Dec. 1999.

A Fuzzy Identity-Based Temporal GIS for the Analysis of Geomorphometry Changes

Myriem Sriti, Remy Thibaud, and Christophe Claramunt

Naval Academy Research Institute, Lanveoc-Poulmic,
BP 600, 29240 Brest Naval, France
{sriti, thibaud, claramunt}@ecole-navale.fr

Abstract. Despite recent progress in the development of temporal Geographical Information Systems (GIS) there is still a lack of methodological integration with geophysical models oriented to the study of Earth changes. This paper introduces a temporal GIS modelling approach which complements a process-based geomorphological experimental apparatus that simulates erosion-sedimentation phenomena over a geological period of time. We combine a field-based with a discrete-based observation of forms and changes at different levels of abstraction. A fuzzy-based model of evolution is introduced and allows for an approximation of changes and processes. State transitions are fuzzy-valued and complemented by a quantitative analysis of change patterns.

1 Introduction

Over the past several years, an important trend in geomorphology studies, where geomorphology is defined as the science of the study of landforms on Earth, has been devoted to the modelling, analysis and interpretation of physical processes such as erosion and sedimentation dynamics [33, 34]. These studies are oriented to the observation and reconstitution of the processes that shape a relief over a geological period of time, in order to derive geophysical models that approximate these evolutions [28]. Due to the inherent difficulty of observing long standing patterns of geomorphology processes, many experimental apparatus have been developed to simulate erosion and sedimentation [43, 14, 13]. Clearly in these approaches, processes are of primary concern, not the terrain regions and landforms that are affected by these physical phenomena. This methodological contrast, that also reveals an incomplete representation of changes, leads us to propose and explore a methodological shift in the modelling approach where the primary emphasis is partly reoriented towards the study of the morphometry, that is, entities and landforms that constitute an observed or simulated system. The dynamics of the system is taken into account by the modelling of the evolution of the relief morphometry, and not only the underlying processes.

We introduce a temporal GIS approach that supports the representation and description of geomorphometrical entities, changes and processes at different levels of abstraction. The aim of the research is to provide a complementary approach to assist scientists in the observation and modelling of either real or experimental systems whose objective is to study the effects of the environment on the geomorphology. The multi-scale character of a geophysical space leads us to consider different levels of

S. Spaccapietra and E. Zimányi (Eds.): Journal on Data Semantics III, LNCS 3534, pp. 81 – 99, 2005.

abstraction and properties of space. From a deductive observation of an erosion model that simulates the evolution of a Digital Elevation Model (DEM) over an experimental apparatus where patterns of changes are observed quantitatively, we introduce a two-step approach where:

- Watersheds and landforms are computerised at successive time steps. Watersheds are elementary hydrology units derived from a field-based analysis of terrain gradients[1] [12, 4]. They form a hierarchical system connected to the direction of flow. Watersheds are delineated using a computational model derived from an experimental erosion-sedimentation apparatus. A spatial constraint is fixed in order to withdraw watersheds whose area is too small to be successfully interpreted (although arbitrary defined this approximation does not have a substantial impact on the development of our modelling approach). The second level of abstraction is that of the landforms that provides a second semantic classification of the relief. These landforms are computerised at a lower level of abstraction using a set of orthogonal geomorphometric primitives derived from a classification of landforms [44]. These mutually exclusive landforms include planes, channels, ridges, passes, peaks and pits. This gives an example of multi-level model where the properties of the relief are analyzed at complementary semantic levels.
- The second component of our modelling approach characterises changes and processes that transform watersheds and landforms. Elementary watershed changes are studied using a fuzzy-based identity relationship that approximates degrees of transformations over time, and the underlying processes that generate them. This fuzzy-based representation is completed by a cross-analysis of landform changes, and observation of diversity and entropy variations over time. Together, this provides a qualitative and quantitative analysis of evolution patterns that complements a process-oriented geophysical modelling of erosion-sedimentation systems.

The reminder of the paper is organized as follows. Section 2 briefly reviews related work in the dynamic modelling of geomorphological systems within GIS. Section 3 introduces the modelling background of our research. Section 4 describes the experimental model of erosion used as a testbed. Section 5 develops the entity-based approach that supports the modelling of watersheds and landforms. Section 6 introduces the fuzzy-based evaluation of watershed changes and processes and the quantitative analysis of landform changes. Finally, Section 7 concludes our paper and outlines further work.

2 Related Work

This section briefly reviews related work in the dynamic modelling of geomorphological systems within GIS. As with earlier attempts, it is worth mentioning the integration of hydrological models and the spatial analysis and visualisation capabili-

---

[1] Following the convincing demonstration made in [29], we consider field-based and vector representations as different structures of space, not conceptual representations which are independent.

ties of the open source GRASS, a raster-based system developed as a non profit software [40, 26]. A significant amount of work has followed in complementing environmental systems with the visualisation functionalities of GIS [6, 8]. In most of these applications, GIS were expected to assist scientists in the visualisation and understanding of geophysical processes, but not the representation of landforms and changes that operate on them. Representing three-dimensional systems as modelling units has been more recently an object of study using voxel-based three-dimensional models [7], extensions of CAD systems [23], and formal topological models [15, 5, 41, 37]. However, those models still do not integrate explicitly the temporal dimension and have not been successfully integrated within GIS and applied to the representation of landforms and geomorphological processes.

A profile-based model of geologic structures is proposed using the GRASS GIS [24], but despite its computational efficiency the approach is still limited to a representation of two-dimensional cross-sections without consideration of changes and evolution. A multi-level approach to the study of soil evolution is developed in [25]. This approach combines quantitative analysis of changes with the analysis of processes at the local level, but there is still no successful integration between the quantitative and qualitative analysis. A fuzzy-based process-based model is developed for the analysis of coastal geomorphological changes [27]. Possible transitions are approximated using fuzzy qualifiers that evaluate successive spatial overlaps between identified entities which are also fuzzy-defined. However, the approach is limited to a single level of abstraction, and no quantitative spatial analysis of evolution complements the approach.

3 Temporal GIS Modelling Background

3.1 Bona Fide vs. Fiat Objects

Within GIS, space has long been represented using either continuous or entity-based models. An intrinsic difficult problem inherent to both models and crucial in the interpretation of environmental phenomena, is the definition of an appropriate level of abstraction. This requires further exegesis at the ontological level in order to fulfil the application requirements. Geographical space can be categorized and modeled using different kinds of entities: proper, structurally integrated entities (e.g. a tree), aggregated objects (e.g. a forest) and intellectually constructed objects [9]. In a related work, a close distinction is also made between *bona fide* and *fiat* objects [38, 39]. *Bona fide* objects materially exist in virtue of intrinsic physical discontinuities (e.g. building, river, lake) while *fiat* objects reflect no intrinsic physical discontinuities but are rather the product of human-drawn boundaries (e.g. mountain, valley). In the field of geomorphology, many of the categories used to qualify terrain characteristics are *fiat* per nature (e.g. watershed, peak, pit, channel).

In our modelling approach, the fiat nature of watersheds is reinforced by the fact that it is not a direct constituent of an observed reality, but a computerised entity derived from the application of hydrological algorithms, and thus not observable as such and that together form a partition of space. Furthermore, watershed delineation is a scale dependent task which is a non straightforward task. This also relates to the distinction between proper and aggregated objects suggested by [9], and the levels of abstraction that fundamentally structure space throughout different hierarchies. The choice of the appropriate level of abstraction is particularly crucial for geomorphological processes as these don't always exhibit a fractal dimension over scale [22]. This brings forward different granularities from the basic particle of land that moves down as a consequence of an erosion-sedimentation process to the evolution of terrain forms and watersheds.

3.2   Quantitative vs. Qualitative Analysis of Changes

The integrated GIS modelling of the spatial and temporal dimensions has been an active research area over the past ten years. The first techniques used for spatio-temporal analysis were quantitative per nature. These rely on the spatial and statistical analysis of remote sensing images, aerial photographs or thematic maps to compare and monitor urban and landscape transformations in a given region of interest [1, 26, 17]. However, quantitative spatial analysis, often based on a continuous representation of space, does not provide any explicit information about the nature of changes and processes at the elementary entity level [11]. There is no representation of temporal networks between successive entity states, and thus the spatio-temporal processes that lead to these changes. This leads GIS research to search for integrated spatio-temporal models for an explicit representation of events, changes and processes within geographical spaces [3, 10, 20, 31, 45, 18]. These spatio-temporal modelling approaches, often qualitative and discrete-based per nature, attempt to integrate time within GIS as a new semantic dimension at the representation, structural and manipulation levels. They lead to spatio-temporal descriptions of geographical entities, changes and processes. To summarise, temporal GISSs can be categorised in two different classes: quantitative views where time and space are modelled as a succession of continuous frames and qualitative where entities are modelled as elementary spatio-temporal objects. This leads to a need for conceptual and processing interoperability between quantitative and qualitative temporal GIS which are crucial for many applications where part of the information is perceived as spatial entities, and the remainder as continuous fields [32].

3.3   Identity

The notion of identity has a substantial influence on the interpretation of the effects of processes on elementary entities and forms [18]. The notion of identity is a property that distinguishes an entity from all others [19, 18]. On the one hand, the evolution of well-defined objects can be tracked throughout their identity and the topological networks constituted. Those evolutions are categorised using taxonomies of spatio-temporal processes where a distinction is made between different forms of evolution (with a conservation of the identity) and mutation (change of identity). On the other hand, *fiat* objects, either interpreted or computerised, are not explicitly identified as such. This leads to a difficulty in analysing evolution as there is no explicit link or network that relate those objects through time. The latter case is exactly the one of the evolution of a partition of watersheds. Although all watersheds are bounded in space, no intrinsic identity property exists, just the spatial properties of the partitions formed by their computation, and the sequence of these partitions through time.

4 Experimental Model

Erosion and sedimentation are complex phenomena, rain and runoff dependent, that result from several elementary processes of detachment, transport and deposition of sediment. In the context of our study, we consider erosion as a geomorphometrical evolution of spatial entities that forms the relief, without taking into account the physical and hydrological factors that generate it (e.g. pluviometry, soil component). A DEM is a fundamental spatial structure widely applied to represent topographic surfaces. A DEM is a regular matrix representation of a continuous variation of relief over space. DEMs enable extraction and identification of continuous properties (e.g. slope, aspect, roughness) and morphological features (i.e. plane, channel, ridge, pass, peak, pit). Over long periods of time, observation and analysis of a series of DEMs give important clues to the way in which a given region of space evolves. As these long standing observations cannot be easily guaranteed, scientists often develop experimental models that replicate real conditions.

The experimental component of our research is based on an analogic model of erosion and sedimentation process whose objective is the study of the influence of natural constraints (e.g. rain, lithology) in relief dynamics. The apparatus consists of a $1m^3$ square box in which a confined dense foggy atmosphere is generated. This enables to simulate rain flow and geomorphologic instabilities above a material obtained by mixing silica granular with water. After rain flow periods, topographic laser digitisation of the obtained surface allows for the modelling of successive DEMs. Experiments give physical information on eroded systems with runoff transportation and topographic incision. This experimental apparatus and the underlying processes are not directly related to a real geomorphological system. However, the erosion phenomena observed are intimately associated to the underlying watersheds. This leads to a scale-dependent relationship between processes and the entities that can be potentially replicated at different levels of abstraction in the analysis of geomorphological processes. Details on the physical analogy and the scaling made between the experimental and natural systems can be found in [13]. The principles of our modelling approach are illustrated using this experimental apparatus and a sample of temporal snapshots that reproduce the effects of the erosion-sedimentation process. We analysed a series of three DEMs, derived from the observation of experimental apparatus evolution, and the partition of watersheds computed (Fig. 1).

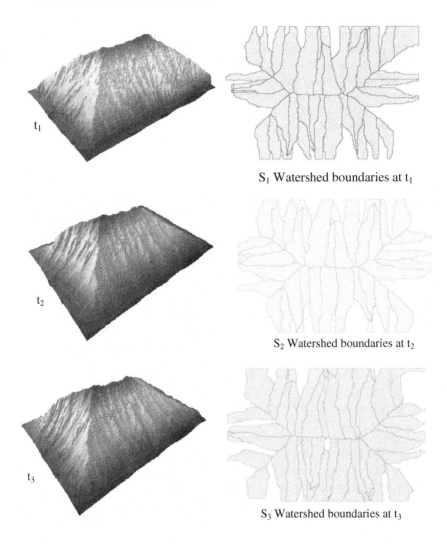

S_1 Watershed boundaries at t_1

S_2 Watershed boundaries at t_2

S_3 Watershed boundaries at t_3

Fig. 1. Temporal series of the erosion model and watershed partitions

5   The Modelling of Watershed and Terrain Forms

Modelling the evolution of watershed and terrain form imply to compare successive snapshots, and cross-analysing the space occupied by the elementary watersheds. Our objective is not to derive a formal model that characterises contraction and dilation processes such as the ones proposed in mathematical morphology and the study of temporal sequences [35], but rather to evaluate to which extent successive watersheds share the same portion of space, or in other words their likeness in representing the evolution of a the same elementary unit. This corresponds to a form of deductive derivation of identity relationships. A generalisation of the approach should generate morphological networks that represent identity relationships through time, and therefore transformations. This analysis should be realized from the level of abstraction of the watershed, to the local level where DEM local characteristics are observed, but with the constraint that too low a level of abstraction might hide observable entities and patterns.

5.1 Watershed-Based Approach

We characterise a watershed by the region of space it occupies, derived geomorphometrical attributes (maximum length, maximum width, area and average slope), and hydrological attributes (stream network, flow direction and accumulation, runoff distances). The first step of the approach consists of the derivation of the watersheds that compose the DEM. This is based on the method implemented in the Hydro Data Model [21] of the Arc Info GIS (© ESRI inc.). Firstly, the computation evaluates the amount of runoff from a rainfall even in a particular area using the SCS curve number method [42]. Secondly, the stream network and the watershed boundaries are delineated using recursive graph functions [2]. As the computation of the watersheds is consistent throughout successive spatial scenes of the experimental model, and under the accuracy limits of the hydrological model, watersheds are considered as crisp objects.

This computation is performed for several DEM snapshots derived from the experimental erosion model. In a mathematical sense, this gives a series of partitions of watersheds over the same region of space. The analysis of a watershed evolution over time relates to the old problem of the evolution of identity [18]. What makes a difference between an evolution that keeps the identity of a given entity in time (the entity evolved but it is still the same) and an evolution that leads to a change of identity (the evolution is significant enough to leads to a change of entity)? This problem is non well-defined and fuzzy by nature. This leads us to model and qualify those phenomena using a fuzzy identity relationship approach. We assume that the identity of a given watershed is maintained to the degree the region of the space where this watershed is located is relatively stable. More formally, the basic principles of our model are as follows.

- Let S denote a subset of a topological space X (i.e. the region of space where the experimental model of erosion is applied). Let $S_1, S_2, ..., S_n$ be an ordered series of partitions of S that denote the evolution of S over time at $t_1 < t_2 < ... < t_n$, respectively. The elements of a given partition S_i are elementary watersheds $W_{i1}, W_{i2}, ..., W_{ip}$ defined as connected regions with no holes.
- Let $W_{ik} \in S_i$ and $W_{jl} \in S_j$ be two watersheds defined at t_i and t_j respectively, with $t_i < t_j$.
- We define a fuzzy identity relationship η of $S_i \times S_j$ as follows:

$$\eta: S_i \times S_j \rightarrow [0, 1] \tag{1}$$

where $\eta(W_{ik}, W_{jl})$ gives the fuzzy membership value that denotes the degree to which W_{jl} of S_j is an evolution of W_{ik} of S_i.

$$\text{with } \eta(W_{ik}, W_{jl}) = \begin{cases} \dfrac{Surf(W_{jl})}{Surf(W_{ik})} \text{ if } W_{ik} \geq W_{jl} \\ \dfrac{Surf(W_{ik})}{Surf(W_{jl})} \text{ if } W_{ik} < W_{jl} \end{cases} \tag{2}$$

where $Surf(W_{ik})$ is an operator that returns the region of space occupied by W_{ik}, and $\dfrac{Surf(W_{jl})}{Surf(W_{ik})}$ the proportion of the region $Surf(W_{jl})$ also occupied by $Surf(W_{ik})$.

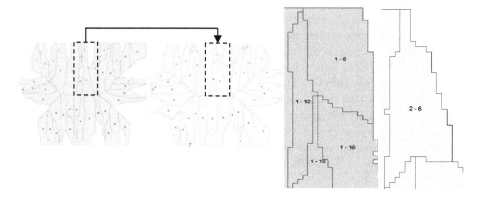

Fig. 2a. Evolution of spatial partition S_1 to S_2

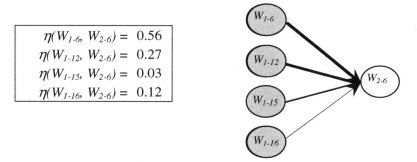

$$\eta(W_{1-6}, W_{2-6}) = 0.56$$
$$\eta(W_{1-12}, W_{2-6}) = 0.27$$
$$\eta(W_{1-15}, W_{2-6}) = 0.03$$
$$\eta(W_{1-16}, W_{2-6}) = 0.12$$

Fig. 2b. Fuzzy identity relation η example

The advantage of this representation is that it gives the identity relationship a degree of membership. When $\eta(W_{ik}, W_{jl})$ tends to 1, the identity relationship between these two watersheds is very high. On the contrary, a value of $\eta(W_{ik}, W_{jl})$ that tends to 0 is likely to denote a non significant identity relationship between these watersheds[2]. In order to mark the difference between evolution and identity, we introduce an operator $Max\text{-}\eta(S_i, W_{jl})$ that returns the watershed(s) of S_i the most related to W_{jl}.

---

[2] This approach can be extended towards a three-dimensional representation using volumes instead of surfaces.

$$Max\text{-}\eta(S_i, W_{jl}) = \{ W_{io} \in S_i \text{ / } \eta(W_{io}, W_{jl}) \geq \eta(W_{is}, W_{jl}), \forall W_{is} \in S_i \} \tag{3}$$

This leads to the identification of a degree of stability $St(W_{jl})$ for a given watershed W_{jl} of S_j with respect to S_i

$$St(S_i, W_{jl}) = \eta(W_{ik}, W_{jl}) \text{ with } W_{ik} \in Max\text{-}\eta(S_i, W_{jl}) \tag{4}$$

High values of $St(S_i, W_{jl})$ tend to reflect stability of W_{jl} since t_i; low values of $St(S_i, W_{jl})$ tend to reflect transformation of W_{jl} since t_i. Fig. 2 illustrates the case of a watershed $W_{2\text{-}6}$ from S_2 and its fuzzy identity relationship with the watersheds of S_1 that share a part of space ($W_{1\text{-}6}$, $W_{1\text{-}12}$, $W_{1\text{-}15}$, $W_{1\text{-}16}$).

In many cases the fuzzy identity relationship is multi-valued as a given watershed is likely to share a part of space with several watersheds of a previous temporal snapshot. We introduce an $\alpha\text{-}Ancst$ operator that returns the "ancestors" over a previous partition S_i of a watershed W_{jl} of S_j, given a fixed threshold α that qualifies the magnitude of the fuzzy identity relationship.

$$\alpha\text{-}Ancst(S_i, W_{jl}) = \{ W_{is} \in S_i \text{ / } \eta(W_{is}, W_{jl}) \geq \alpha \} \tag{5}$$

Applied to the case presented in Fig. 2 this gives for example $0.5\text{-}Ancst(S_1, W_{2\text{-}6}) = \{W_{1\text{-}6}\}$ with $\alpha = 0.5$.

This watershed-based approach allows for an observation of the geomorphological changes of an erosive relief. The proposed model helps to make the difference between degrees of stability and evolution for a given watershed over a period of time. The periods of time considered in our experiment are those given by the numerical representations of the successive erosion model evolutions. Generalization of the approach to the whole region of study allows for an identification of patterns of stability versus evolution using the stability operator.

5.2 Landform-Based Approach

At a lower level of abstraction, observed entities are salient features that participate in the irregularities of the DEM (i.e. planes, channels, ridges, passes, peaks, pits). We retain a grid-based approach that locally computes terrain features using a neighborhood analysis [44]. This computation is based on the second derivative expressions given in Table 1 (cf. p. 112-118 in [44]).

These landform classes are widely used in geomorphometry when defining local surface forms. They are characteristic of any surface independently of the process that affects it. The name of these landforms suggests a geomorphological interpretation, but they may be unambiguously described in term of rates of change of three orthogonal components [44].

Despite the fact that these landforms can be analysed at different levels of resolution [16], we retain the 3*3 matrix suggested in [44] and described in Table 1 for a given level of abstraction. This is acceptable for the objectives of a quantitative analysis.

Table 1. From Wood (1996)

Landform	Feature name	Derivative expression	Description
	Peak	$\dfrac{\delta^2 z}{\delta x^2} > 0, \dfrac{\delta^2 z}{\delta y^2} > 0$	Point that lies on a local convexity in all directions (all neighbours lower).
	Ridge	$\dfrac{\delta^2 z}{\delta x^2} > 0, \dfrac{\delta^2 z}{\delta y^2} = 0$	Point that lies on a local convexity that is orthogonal to a line with no convexity/concavity
	Pass	$\dfrac{\delta^2 z}{\delta x^2} > 0, \dfrac{\delta^2 z}{\delta y^2} < 0$	Point that lies on a local convexity that is orthogonal to a local concavity
	Plane	$\dfrac{\delta^2 z}{\delta x^2} = 0, \dfrac{\delta^2 z}{\delta y^2} = 0$	Point that do not lie any surface concavity or convexity
	Channel	$\dfrac{\delta^2 z}{\delta x^2} < 0, \dfrac{\delta^2 z}{\delta y^2} = 0$	Point that lies on a local concavity that is orthogonal to a line with no concavity/convexity
	Pit	$\dfrac{\delta^2 z}{\delta x^2} < 0, \dfrac{\delta^2 z}{\delta y^2} < 0$	Point that lies on a local concavity in all directions (all neighbours higher).

6 Watershed and Terrain Forms Evolution

As in space, the temporal dimension brings the same kind of duality where a distinction can be made between *bona fide* (e.g. building destruction) and *fiat* processes (e.g. erosion process). The nature of geomorphological processes is continuous although here observed and computerised at successive temporal timestamps, thus giving a sequence of temporal snapshots of the experimental region of study. These temporal snapshots give a quantitative estimation of the effects of the erosion process at different time stamps. The influence of the temporal granularity on these evaluations is left to further work.

We introduce a two-step approach of evolution modelling. Firstly, an entity-based observation of transformations models watershed changes and the underlying spatial processes that generate them. Secondly, a quantitative analysis of changes evaluates to which degree landforms do evolve at the local level.

6.1 Watershed-Based Changes

A first class of process, or rather an absence of process, is the stability of a watershed. We define an α-*Stability* as follows

- α-*Stability*: a watershed W_{jl} of S_j is considered in α-*Stability* since t_i of S_i iff $St(S_i, W_{jl}) \geq \alpha$.

We do not consider appearance and disappearance processes as we assume that watersheds are either relatively stable or transformed. A watershed α-*Deformation*, considered in a two-dimensional sense, reflects an evolution significant enough to denote a deformation over a given threshold α, it is defined as

- *(1-α)-Deformation*: a watershed W_{jl} of S_j is considered in *(1-α)-Deformation* since t_i of S_i iff $St(S_i, W_{jl}) < \alpha$.

Given a watershed W_{jl} of S_j, α and S_i, it is immediate to note that W_{jl} is either in α-*Stability* or *(1-α)-Deformation* since t_i of S_i. Quantitatively, an *(1-α)-Deformation* denotes a combined change of geomorphological properties such as area, length and width of a given watershed. Fig. 3 illustrates the trends revealed by observation of *(1-α)-Deformation* over α on S_1 - S_2 and S_2 - S_3. Fig. 3 shows that these two transitions have similar numbers of lowly and highly deformed watersheds. Transition from S_1 to S_2 reveals a higher number of middle range deformations but the deformation process is more important between S_2 and S_3. Two factors can explain these trends: either the physical environment that generates the erosion-sedimentation process has been modified, or the relief response to the process has changed. However, this analysis is left to further study and interpretation from the geomorphologist.

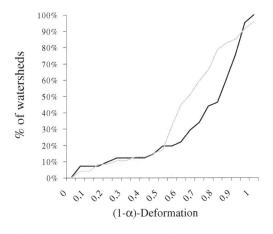

Fig. 3. (1-α)-Deformation over α on S_1-S_2 (light grey) and S_2-S_3 (dark grey)

A specific case concerns the spatial transitions that involve several watersheds. These correspond to three fundamental cases: β-*Union* of n watersheds into a single one, β-*Split* of one watershed towards m watersheds, β-*Reallocation* of n watersheds into m watersheds. These processes are defined as follows:

- β-*Union*: one watershed W_{jl} of S_j is the β-*Union* of W_{i1}, W_{i2}, ..., W_{in} of S_i iff W_{i1}, W_{i2}, ..., $W_{in} \in \beta$-*Ancst*(S_i, W_{jl}).
- β-*Split*: one watershed W_{il} of S_i is splitt towards W_{j1}, W_{j2}, ..., W_{jm} of S_j iff $W_{il} \in \beta$-*Ancst*$(S_i, W_{j1}) \wedge W_{il} \in \beta$-*Ancst*$(S_i, W_{j2}) \wedge ... \wedge W_{il} \in \beta$-*Ancst*$(S_i, W_{jm})$.
- β-*Reallocation*: n watersheds W_{i1}, W_{i2}, ..., W_{in} that form a subset S_{il} of S_i are reallocated towards W_{j1}, W_{j2}, ..., W_{jm} that form a subset S_{jl} of S_j iff $(\forall\ W_{ik} \in S_{il}\ \exists\ W_{jl} \in S_{jl}\ /\ W_{ik} \in \beta$-*Ancst*$(S_i, W_{jl})) \wedge (\forall\ W_{jl} \in S_{jl}\ \exists\ W_{ik} \in S_{il}\ /\ W_{ik} \in \beta$-*Ancst*$(S_i, W_{jl}))$.

These processes give qualitative indicators that characterise watershed evolution over the erosion-sedimentation model. The β coefficient acts as a flexible parameter that supports derivation of evolution networks that reveal different patterns of union, split and reallocation processes. Reallocation processes are of particular interest as

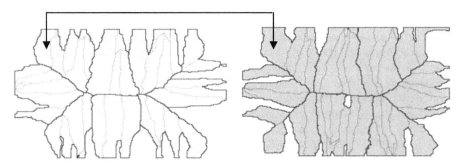

Fig. 4a. β-*Reallocation* of watersheds from S_2 to S_3 with $\beta = 0.2$

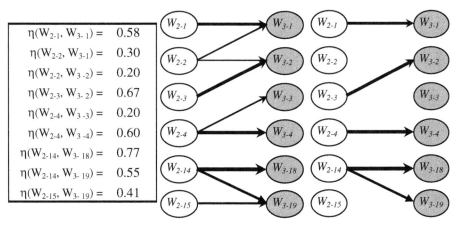

$\eta(W_{2\text{-}1}, W_{3\text{-}1}) =$	0.58
$\eta(W_{2\text{-}2}, W_{3\text{-}1}) =$	0.30
$\eta(W_{2\text{-}2}, W_{3\text{-}2}) =$	0.20
$\eta(W_{2\text{-}3}, W_{3\text{-}2}) =$	0.67
$\eta(W_{2\text{-}4}, W_{3\text{-}3}) =$	0.20
$\eta(W_{2\text{-}4}, W_{3\text{-}4}) =$	0.60
$\eta(W_{2\text{-}14}, W_{3\text{-}18}) =$	0.77
$\eta(W_{2\text{-}14}, W_{3\text{-}19}) =$	0.55
$\eta(W_{2\text{-}15}, W_{3\text{-}19}) =$	0.41

Fig. 4b. Fuzzy identity relationships

Fig. 4c. Reallocation example $\beta = 0.2$

Fig. 4d. Reallocation example $\beta = 0.5$

they underline, for a given β value, the region of space which is "internally" spatially restructured, thus denoting a relative stability over time of the region covered by the union of the watersheds involved. This also has the advantage of determining a scale where regions are relatively stable.

Fig. 4 illustrates how reallocations evaluated with an appropriate β (β=0.2 in this case) clearly delineates homogeneous regions in space and time, that is, regions which are internally spatially redistributed (Fig. 4a – regions delineated in bold). On the contrary Fig. 4d shows how a higher value of the coefficient does not exhibit significant fuzzy identity relationships. An example of reallocated region is illustrated in Fig. 4a, 4b and 4c. This reveals a significant property of the modelling approach: from a lower level of abstraction retained for the watershed delineation (i.e. DEM high resolution), the fuzzy identity relationship allows for a derivation of homogeneous watersheds in space and time at a higher level of abstraction.

The quantitative analysis of watershed changes is developed through a comparison of geomorphomological 2D and 2.5D attributes. The identity relations between the watersheds at different times have been identified by an application of the fuzzy-based relation developed in Section 5 (e.g. 40 watersheds identified at S_1, 41 at S_2, 47 at S_3).

We introduce a function λ that evaluates the degree of change of form between a given watershed W_{jl} and one of its α-ancestors

$$\lambda(W_{ik}, W_{jl}) = \frac{Length^2(W_{jl})}{Area(W_{jl})} - \frac{Length^2(W_{ik})}{Area(W_{ik})} \qquad (6)$$

with $W_{ik} \in \alpha\text{-}Ancst(S_i, W_{jl})$, where $Length(W_{jl})$ returns the maximum length of W_{jl} and $Area(W_{jl})$ the area of W_{jl}.

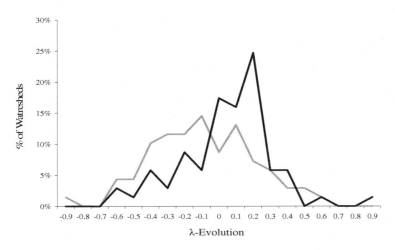

Fig. 5a. Distribution of watersheds over λ-evolutions between S_1 and S_2 (light grey) and S_2 and S_3 (dark grey)

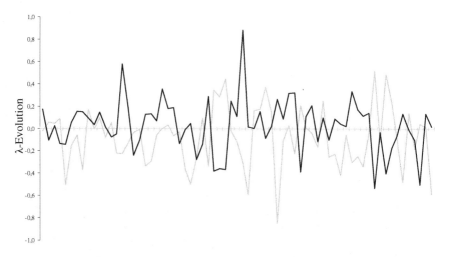

Fig. 5b. Watershed λ-evolutions between S_1 and S_2 (light grey), and between S_2 and S_3 (dark grey)

High positive values of $\lambda(W_{ik}, W_{jl})$ denote a lengthening process, low negative values of $\lambda(W_{ik}, W_{jl})$ denote a widening process. Values that tend to 0 denote a process where W_{ik} and W_{jl} have the same form despite the changes that happened between t_i and t_j. Fig. 5a and 5b illustrate the patterns exhibited by the function λ over the erosion process. Fig. 5a shows that the lengthening process is important between S_2 and S_3, while not significant between S_1 and S_2 where the proportion of watersheds that widen is more significant. Fig. 5b confirms this trend as almost the half of the watersheds first widen between S_1 and S_2, and then lengthen between S_2 and S_3.

6.2 Landforms-Based Changes

The quantitative observation of landform and watershed changes are realised in several steps. At the local level, a general trend is given by the evolution of landform diversity and dominance indexes over the successive temporal snapshots. A cross analysis of landform changes evaluates landform transformation patterns. The diversity index indicates the degree to which a given number of landform classes are represented on a map in equal proportion [36, 30]. It is given by

$$H = -\sum_{k=1}^{m} P_k \ln(P_k) \tag{7}$$

where m is the number of landform classes, P_k the relative area of each one.

The dominance index D is normalised to a range of values from 0 to 1 and measures the extent to which one or few features dominate the terrain. It is given by

$$D = \left(\frac{H_{max} - H}{H_{max}} \right)$$

(8)

where H_{max} is the maximum diversity.

$$H_{max} = \ln(m)$$

(9)

Table 2. Cross - tabulation matrices between S_1 and S_2, and between S_2 and S_3

S_2 \ S_1	Peak (24.22%)	Ridge (4.11%)	Pass (44.76%)	Plane (0.29%)	Channel (3.76%)	Pit (22.86%)
Peak (22.67%)	32.91%	3.56%	44.68%	0.12%	2.27%	16.46%
Ridge (5.00%)	22.85%	15.88%	34.90%	1.38%	11.24%	13.75%
Pass (45.65%)	24.24%	2.92%	46.67%	0.11%	2.68%	23.38%
Plane (0.28%)	6.82%	36.36%	19.32%	10.23%	20.45%	6.82%
Channel (4.39%)	16.45%	11.09%	34.76%	2.22%	15.74%	19.74%
Pit (22.02%)	17.33%	2.67%	45.43%	0.09%	3.24%	31.25%

S_3 \ S_2	Peak (25.65%)	Ridge (3.51%)	Pass (45.41%)	Plane (0.21%)	Channel (3.30%)	Pit (21.92%)
Peak (24.22%)	29.22%	2.71%	45.02%	0.03%	2.03%	20.99%
Ridge (4.11%)	20.63%	7.57%	32.16%	1.68%	11.76%	16.20%
Pass (44.76%)	26.09%	2.51%	47.11%	0.06%	2.39%	21.84%
Plane (0.29%)	8.60%	30.11%	18.28%	7.53%	29.03%	6.45%
Channel (3.76%)	16.69%	11.35%	34.97%	1.75%	17.45%	17.78%
Pit (22.86%)	23.58%	2.14%	46.94%	0.08%	2.27%	24.99%

Low values of diversity indicate that the terrain is composed of several landform classes represented approximately in equal proportion. Conversely, high values reveal that the terrain is dominated by one or only a few landform classes. The dominance indices measured on S_1, S_2 and S_3 give respectively $D_1 = 0.57$, $D_2 = 0.57$ and $D_3 = 0.56$, thus a very stable and relatively high diversity. Despite this diversity stability, the cross-analysis presented in Table 2 shows that a lot of landform transitions do happen between S_1 and S_2, and S_2 to S_3.

The application of Wood's landform classification to the experimental apparatus illustrated in Fig. 1 reveals a high proportion of peaks, pits and passes (cf. Table 2). This is characteristic of the cumulated impact of the erosion model and the local heterogeneity of the DEM.

Overall, the entity-based and landform-based analysis of changes provide complementary views on the geomorphological processes. They illustrate the potential of our approach (however, no specific conclusions can be made on the experiments as the apparatus is not related to a specific scale).

7 Conclusion

Despite significant progress in the development of temporal GISs, there is still a need for the development of multi-dimensional spatio-temporal models suited to the complexity of environmental applications where the identification of entities, changes and processes are themselves part of the scientific process. Based on a deductive observation and quantification of an erosion phenomenon, we introduce a two step modelling approach that combines qualitative and quantitative analysis of changes. At the entity level, qualitative per nature, watershed changes are tracked using a fuzzy-based approach that qualifies degrees of stability, different forms of evolutions, and homogeneous regions over time. Quantitative evaluation of landform distributions and transformations complement the approach by providing global indicators on erosion-sedimentation phenomena.

Our approach provides a complementary view on the evolution of watershed units and landforms which are intimately linked to these processes. The fuzzy component of the method favours flexibility and different degrees of characterisation of changes, thus giving the flexibility which is compatible with the complexity and the stochastic component of geomorphological processes. The model is illustrated by an experimental erosion model that simulates morphometric changes in a region of study over a long period of time. The case study shows the real potential of our approach for geomorphology studies. An aspect of the watershed-based approach that still needs to be studied is the influence of the temporal granularity on the evolutions derived from the fuzzy-based modelling. This relates to the old problem of the dependence between phenomena and its levels of observation in time, but also in space. There is also still a need to further explore a level of abstraction on the landform observation compatible to the one identified for the computation and aggregation of homogeneous watershed regions.

The principles of our fuzzy-based modelling approach can be applied to other discrete geographical phenomena where there is no explicit way of tracking entity identity over time, and where different levels of abstraction are possibly required for the analysis of changes. The deductive representation of changes supported by the fuzzy based model generates a sort of semantic network over time whose links are spatial per nature in the case of geomorphology. However, the structure and the topology of such a temporal and semantic network can potentially represent any relationship over time, thus giving a larger generality to the model. Further work concerns the extension of the model to a further integration of the qualitative and quantitative dimensions of the model in the spatial and temporal dimensions. We also plan to extend the fuzzy-based reasoning mechanisms to integrate the third spatial dimension.

References

1. Anselin, L. and Getis, A.: Spatial statistical analysis and geographic information systems. The Annals of Regional Science, 26 (1992) 19-33
2. Band, L. E.: Spatial hydrography and landforms. In P. A. Longley, M. F. Goodchild, D. J. Maguire and D. W. Rhind (eds.), Geographical Information Systems, 2nd edition, Wiley, London, (1999) 527-542
3. Beller, A.: Spatial-temporal events in GIS. In Proceedings of GIS/LIS 91, 57(4) (1991) 407-411
4. Beven, K. J. and Wood, E. F.: Catchment geomorphology and the dynamics of runoff contributing areas. Journal of Hydrology, 65 (1983) 139-158
5. Billen, R. and Zlatanova, S.: 3D spatial relationships model: a useful concept for 2D cadastre? Computers, Environment and Urban Systems, 27 (2003) 411-425
6. Bishop, I. D. and Karadaglis, C.: Linking modelling and visualisation for natural resources management. Environmental and Planning B: Planning and Design, 24(3) (1997) 345-358
7. Breunig, M.: An approach to the integration of spatial data and systems for a 3D geoinformation system. Computer and Geosciences, 25 (1999) 39-48
8. Burrough, P. A.: Dynamic modelling and geocomputation. In P. Longley, S. Brooks, R. McDonnell and McMillan (eds.). Geocomputation: A Primer, John Wiley & Sons, New York (1998) 165-191
9. Chapman, G. P.: Human and Environmental Systems: A Geographer's Appraisal, London, Academic Press (1977)
10. Claramunt, C. and Thériault, M.: Managing time in GIS: An event-oriented approach. In J. Clifford and A. Tuzhilin (eds.), Recent Advances on Temporal Databases, Springer-Verlag, Zurich (1995) 21-43
11. Claramunt, C. and Thériault, M.: Toward semantics for modelling spatio-temporal processes within GIS. In M. J. Kraak and M. Molenaar (eds.), Advances in GIS Research, Delft, Taylor & Francis (1996) 47-63
12. Clarke, R. T.: A review of some mathematical models used in hydrology, with observations on their calibration and use. Journal of Hydrology, 19 (1973) 1-20
13. Crave, A., Lague, D., Davy, P. Kermarrec, P. and Sokoutis, J.: Analog modelling of relief dynamic. Physics and Chemistry of the Earth , 25 (2000) 549-553
14. Czirok, A., Somfai, E. and Vicsek, T.: Experimental evidence for sel-affine roughning in a micro-model of geomorphological evolution. Physical review Letters, 71 (1993) 2154-2157
15. De la Losa, A. and Cervelle, B.: 3D topological modelling and visualisation for 3D GIS. Computers & Graphics, 23 (1999) 469-478
16. Fisher, P., Wood, J., and Cheng, T.: Where is helvellyn? Fuzziness of multi-scale landscape morphometry. Transactions of the Institute of British Geographers, 29 (2004) 106-128

17. Goodchild, M. F., Steyaert, L. T. and Parks, B. O. (eds.): GIS and Environmental Modelling: Progress and Research Issues, Fort Collins, GIS World Books (1996)
18. Hornsby, K. and Egenhofer, M.: Qualitative representation of change. In A.U. Frank and D. Mark (eds.), Spatial Information Theory (COSIT'97), Springer-Verlag (1997) 15-33
19. Koshafian, S. and Copeland, G.: Object identity. SIGPLAN Notices, 21 (1986) 406-416
20. Langran, G.: States, events and evidence : the principle entities of a temporal GIS. In Proceedings of GIS/LIS'92 (1992) 416-425
21. Maidment, D. R., and Djokic, D. (eds.): Hydrologic and Hydraulic Modeling Support with GIS, ESRI Press, Redlands CA (2000) 216 p
22. Mark, D. M. and Aronson, P. B.: Scale-dependent fractal dimensions of topographic surfaces: An empirical investigation with applications in geomorphology and computer mapping. Mathematical Geology, 16(7) (1984) 671-683
23. Marschallinger, R.: A voxel visualisation and analysis system based on Autocad. Computer and Geosciences, 22 (1996) 379-386
24. Matsumoto, S., Raghavan, V., Yonezawa, G., Nemoto, T. and Shiono, K.: Construction and visualisation of a three dimensional geologic model using GRASS GIS. Transactions and GIS, 8(2) (2004) 211-223
25. Mendonça, L. and Claramunt, C.: An integrated landscape and local analysis of land cover evolution in an alluvial zone. Computer Environment and Urban Systems, Pergamon (Pub.), 25(6) (2001) 557-577
26. Mitasova, H., Mitas, L., Brown, W., Gerdes, D., Kosinovsky, I. and Baker, T., Modeling spatially and temporally distributed phenomena: new methods and tools for GRASS GIS. International Journal of GIS, 9(4) (1995) 433-446
27. Molenaar, M. and Cheng, T.: Fuzzy spatial objects and their dynamics. ISPRS Journal of Photogrammetry and Remote Sensing, 55 (2000) 164-175
28. Montgomery, D. R., Balco, G., and Willet, S. D.: Climate, tectonics and the morphology of the Andes. Geology, 29 (2001) 579-582
29. Nunes, J.: Geographic space as a set of concrete geographical entities. In D. M. Mark and A. U. Frank (eds.), Cognitive and Linguistic Aspect of Geographical Space, Kluwer Academic Publishers, (1991) 9-33
30. O'Neill, R.V., Krummel, J.R., Gardner, R.H., Sugihara, G., Jackson, B., De Angelis, D.L., Milne, B.T., Turner, M.G., Zygmunt, B., Christensen, S.W., Dale, V.H., and Graham, R.L.: Indices of landscape pattern. Landscape Ecology, 1(13) (1988) 153-162
31. Peuquet, D. J.: It's about time: a conceptual framework for the representation of temporal dynamics in geographic information systems. In Annals of the Association of American Geographers, 84(3) (1994) 441-461
32. Peuquet, D. J.: Making space for time: Issues in space-time data representations. Geoinformatica, 5(1) (2001) 11-32
33. Pike, R. J.: A Bibliography of Geomorphology, United States Geological Survey Open File Report 93-262-1, Menlo Park, CA (1993)
34. Rhoads, B. L. and Thorn, C. E.: The scientific nature of geomorphology. In Proceedings of the 27th Binghantom Symposium in Geomorphology, Chichester, New York, John Wiley & Sons (1996)
35. Serra, J.: Image Analysis and Mathematical Morphology, Academic Press, London, (1982)
36. Shannon, C. E. and Weaver, W.: The Mathematical Theory of Communication, Urbana IL, University of Illinois Press (1949)
37. Shi, W., Yang, B. and Li, Q.: An object-oriented data model for complex objects in three-dimensional GIS. International Journal of GIS, 17(5) (2003) 411-430

38. Smith, B.: Fiat Objects, in N. Guarino, L. Vieu and S. Pribbenow (eds.), Parts and wholes: conceptual part-whole relations and formal mereology. In Proceedings of the 11th European Conference on Artificial Intelligence, Amsterdam (1994) 15-23

39. Smith, B. and Mark, D. M.: Do mountains exist? Towards an ontology of landforms. Environment and Planning B: Planning and Design, 30(3) (2003), 411–427

40. Srinivisan, , R. and Arnold, J. G.: Integration of a basin-scale water quality model with GIS. Water Resources Bulletin, 30(3) (1994) 453-462

41. Tse, R. and Gold, C.: A proposed connectivity-based model for a 3-D cadastre. Computers, Environment and Urban Systems, 27 (2003) 427-445

42. USDA-SCSNational Engineering Handbook, Section 4 – Hydrology, Washington D.C., USDA-SCS (1985)

43. Wittmann, R., Kautzky, T., Hübler, A. and Lücher, E.: A simple experiment for the examination of dendritic river systems. Naturwissenschaften, 78 (1991) 23-25

44. Wood, J.: The Geomorphological Characterisation of Digital Elevation Models, Unpublished PhD report, University of Leicester, UK (1996)

45. Worboys, M.: A unified model for spatial and temporal information. The Computer Journal, 37(1) (1994) 26-34

Interoperability for GIS Document Management in Environmental Planning

Gilberto Zonta Pastorello Jr., Claudia Bauzer Medeiros,
Silvania Maria de Resende, and Henrique Aparecido da Rocha

Laboratory of Information Systems – Institute of Computing,
University of Campinas – CP6176, 13081-970 – Campinas, SP, Brazil
{gilberto, silvania, henrique}@lis.ic.unicamp.br
cmbm@ic.unicamp.br

Abstract. Environmental planning requires constant tracing and revision of activities. Planners must be provided with appropriate documentation tools to aid communication among them and support plan enactment, revision and evolution. Moreover, planners often work in distinct institutions, thus these supporting tools must interoperate in distributed environments and in a semantically coherent fashion. Since semantics are strongly related to use, documentation also enhances the ways in which users can cooperate. The emergence of the Semantic Web created the need for documenting Web data and processes, using specific standards. This paper addresses this problem, for two issues: (1) ways of documenting planning processes, in three different aspects: *what* was done, *how* it was done and *why* it was done that way; and (2) a framework that supports the management of those documents using Semantic Web standards.

1 Introduction

Environmental planning covers many aspects and geographical scales, ranging from a city section to the global level. It is a continuous process that requires to constantly monitor the region under study. Multidisciplinarity and dependence on cooperative work are characteristics of environmental planning activities.

During the development of environmental plans many steps are carried out. Among them can be singled out: (1) identification of problems to be considered in a given geographic area – the "diagnosis"; (2) development of strategies to solve or minimize these problems at short, medium and long term – the "plan"; (3) Implementation of the chosen strategies – plan "execution"; (4) plan revision and maintenance – "follow-up". This process is strongly based on using geographical data and Geographical Information Systems (GIS).

Steps (1) and (2) are backed up by two kinds of document sets:

- A set of maps and related descriptive data which detail the characteristics of the studied region. Maps usually portray two types of situation: the current situation, which is the input to the planning activity; and the possible outcomes of plan execution (the desired final state);

S. Spaccapietra and E. Zimányi (Eds.): Journal on Data Semantics III, LNCS 3534, pp. 100–124, 2005.

– A set of directives which specify how to achieve the planning goals, enacting them using the maps as background.

Plan execution (3) is the implementation of the directives. At each stage there are several alternatives that should be discussed and revised by teams, considering, for instance, options on preservation or recovery of environmental resources to be balanced against economic exploitation constraints.

This process requires detailed documentation, but there is a lack of tools to support document management. As a consequence, if a similar problem occurs in another region, it is necessary to start from scratch. This hampers plan modification and detection of methodological errors. Documentation is also important for communication among designers, in order to aid plan maintenance and evolution. As the planning process grows in complexity, more people and technologies must be involved, augmenting the need for documentation. Moreover, documentation provides information on the use of given datasets, and the context in which they are used. Thus, it provides additional semantics to a given planning procedure.

Yet another factor to consider is the fact that Spatial Decision Support is moving from a closed, tightly controlled computational environment to an open, Web-based context. This brings up new research and development challenges. Web GIS can no longer be seen only under the perspective of GIS accessed via the Web. They must also consider that their users and data are distributed all over the world. Thus, the Web has created not only the need for GIS distribution and interoperability but also requires offering domain experts easy means of publishing and accessing distributed resources and documents.

This paper presents a computational framework to support cooperative environmental planning activities on the Semantic Web. This framework is centered on the notion that documentation is a key issue in fostering collaboration and reuse and attaching more semantics to data and procedures. In this context, documentation should describe not only the data used – e.g. a region's geophysical and economic context – but also the planning process itself. Based on these observations, the proposed framework supports management of three main kinds of documents on the Web: *what* was done, *how* the plan was produced, and *why* the plan was developed along given lines.

Part of the framework has already been implemented at the University of Campinas, where these documents have proven to be useful in a local context. This implementation led to the Decision Support System named WOODSS (Workflow-based spatial Decision Support System), see [21, 35]. It has been used to test and validate ideas related to environmental planning support and associated documents [33].

However, in order to support cooperation across the Web, semantics and interoperability issues must be considered. Answering this need, this paper extends the documentation paradigm to the Semantic Web in two ways. First, it adopts XML to represent these documents, thereby providing the basis for interoperability. Furthermore, it discusses the use of existing domain ontologies as the means to attach further semantics to documents, data and planning processes, levering cooperation and automatic execution of processes on the Web. We furthermore adopt Web Services for framework implementation. The result is a step towards fully interoperable Spatial Decision Support Systems.

The remainder of this paper is organized as follows. Section 2 presents some basic concepts and related work. Sections 3 and 4 specify the three kinds of documents, detailing internal database and Semantic Web representations. Section 5 presents the WOODSS system and implementation issues. Section 6 shows an application example. Finally, Section 7 presents conclusions and ongoing work.

2   Related Work and Basic Concepts

The main concerns in our work involve documentation of planning procedures and the Semantic Web. Related work is thus centered on these issues.

Documentation adopted by environmental planning experts is highly unstructured. It is usually maintained in very large textual files. Automated support for such documentation is limited to text processing tools. Also, domain experts largely ignore Computer Science advances in this area. Consequently, there are few studies on document management for environmental activities.

Our research takes as starting point one of the few works that deals with documentation within a geographic context, viz. [30]. This work proposes the management of *What, How* and *Why* documents associated with the changes occurring in a spatiotemporal database, to support a better understanding of the evolution of geographic phenomena in the context of urban development applications. Documentation and spatial objects are managed jointly in a single database, in order to document change reasons, procedures and originators.

Our documentation goals, as will be seen, require a finer grain of detail, due to the particularities of environmental activities. Specifically, our *What* documents consist of metadata as well as additional data stored in hypertext/hypermedia graphs. Furthermore, like [30], workflows are used to store *How* documents, and design rationale for *Why*. There follows a short survey on related work in issues for each of these documentation choices.

2.1   Hypermedia and Metadata

Hypermedia represents an approach for management of information where data are stored in a network of nodes connected by links. A node represents a concept or idea and contains some multimedia data, such as text, graphics, video or images. Links represent relationships between nodes. The content of a node is presented by activation of links.

Hypermedia technology is used in applications that manage dynamic documents as in digital libraries [28] or at the Web. Also it can be used in other contexts, e.g., version control [19] and integration of heterogeneous software development environments [2]. For a formal representation and comparison of different hypermedia data models see [46].

The Dexter model [17] is a widely adopted hypermedia reference model, where a hyperdocument consists of a set of *components*. A component includes a *con-*

tents specification, a general-purpose set of *attributes*, a *presentation specification* and a set of *anchors*. A component can be an *atom*, a *link* or a *composite*. The atomic component represents the hypermedia 'node' abstraction, containing generic data. Links are entities that represent relationships between components. The contents of a link component is a list of *specifiers*, each including a presentation specification as well as component and anchor identifiers.

DHM [26] is an object oriented open hypermedia system based on Dexter. Its data model extends Dexter's links, anchors and components/compositions. The model supports dangling links – links having zero or one endpoint – and anchoring is extended to include a distinction between marked and unmarked anchors.

Other models in the literature extend Dexter to include, for instance, adaptive techniques or semantic connectors. For our purpose, however, it suffices to use Dexter's basic model and some extensions proposed in DHM. Hypermedia serves as a basis for storing *What* documents, enhanced with metadata.

Metadata, in the sense of data that describe data, are useful in many contexts – documentation, semantics and support for data retrieval. In the GIS context, metadata are classified in three levels [10]: description of the studied domain; characteristics of exchanged data; and characteristics of the geographic information. Several metadata standards have been proposed for storing and exchanging geographic data. WOODSS' metadata [34] complement What-documentation and are based on the FGDC's Content Standard for Digital Geospatial Metadata (CSDGM) proposal [13]. They contain information on spatial and temporal characteristics, as well as lineage and quality information.

2.2   Design Rationale

During a design process, many alternatives can be adopted. Designers need to analyze each option and choose the more suitable one according to goals to be reached. Design rationale (DR) is an artificial intelligence technique that supports a formal representation of the reasons behind decisions taken in a design process. It allows keeping track of assumptions made during this process, and the discussions conducted within a design team – and sometimes across teams – to arrive at a given solution. DR is object of research mainly in Artificial Intelligence [6], Software Engineering [18] and Human-Computer Interfaces [24]. Our work uses an extension of these techniques for creating *Why-documents* in environmental planning activities.

DR usually adopts semi-formal methods based on directed graphs, where edges and nodes acquire specific semantic meaning. All these models have a set of basic elements that formalize the discussions around a given project – the questions posed, the alternatives that are raised in response to questions and the arguments for and against the alternatives. These elements, which can be interlinked, are represented in IBIS (Issue-Based Information Systems) [11], a pioneer effort to formalize DR, through the entities *Issues*, *Positions* and *Arguments*. The links can be of eight kinds and they have intuitive meaning. For example, a Position <Responds-to> an Issue; Arguments must be linked to their Positions with either <Supports> or <Objects-to> links; and so on.

Other models include PHI [16] that uses similar concepts to IBIS, and Design Space Analyses (DSA) [23]. Proteus [25] is a model for documenting and managing the rationale of software design. DR techniques have been used in other contexts, e. g. support for design reuse and collaborative design in design engineering projets [50].

2.3   Workflows and Scientific Workflows

A *workflow* denotes the controlled execution of multiple tasks in an environment of distributed processing elements. It can be defined as a set of tasks involved in a procedure along with their interdependencies, inputs and outputs. Each task is called an *activity*, which is a unit of work and can be executed by one or more agents, in a given *role*. An *agent* is a person or software component able to execute one or more activities.

Traditionally, workflows have been used for total or partial automation of business processes. *Scientific workflows* [37, 43] allow documenting and specifying scientific experiments and procedures. Scientific work documentation requires special treatment because it is characterized by a great degree of flexibility and presents a much higher amount of uncertainty and exceptions than business work. Scientific workflows extend business workflows functionality supporting the following aspects: *incompleteness*; *partial re-use*; *abandon/rewind and dynamic modification*; *tracing of invalid processes*; *specification from case*. For a description of these aspects see [1].

In business applications, the main motivation for introducing workflow management is the desire to "re-engineer" work to enhance efficiency. The motivation for workflow management in scientific applications, additionally, is to help to control experiments, and to make available to scientific users the information on how experiments were conducted [35]. A recent trend concerns the use of workflows across the Web, to support cooperative work organization (e.g., the special issue on internet-based workflows in [31], or the work of [8] on coordinating communication among workflows).

Environmental planning activities have the same peculiarities of scientific work procedures. Thus, we adopt the scientific workflow paradigm to document *How* these activities are performed. Examples of this kind of use are [5, 7, 8, 20, 31, 32], involving geospatial data for e-government, in situation of emergency planning and environmental disaster management.

2.4   Semantic Web Related Efforts and Standards

Our choices for document representation favor flexibility in document construction and ease in document exchange, by following specific standards. Such characteristics are important when it comes to cooperative processes, and become essential when we consider our ultimate goal, that is interoperability and reuse in the Web. The Semantic Web is being proposed as an evolution of today's Web to make the information available on the Web easily usable, with the aid of automatic tools. The World Wide Web Consortium is the association that leads the standards efforts on the Web and Semantic Web [42].

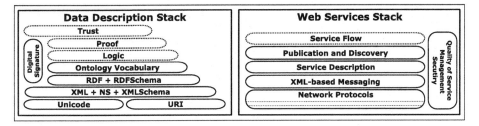

Fig. 1. The Semantic Web and Web Services Standards: Data and Services Description

The conceptual separation between data and services induces an implementation for the Semantic Web. On one side there are the data that should be semantically understood in the same way wherever they are used. On the other, pieces of software should provide a satisfactory degree of automation when handling these data. Such pieces of software often are Web Services. Figure 1 shows the proposal for data and services standards structure, portrayed in layers, where each layer supports the construction of the ones on top of it. Layers within dotted boxes do not yet have consensual standards.

At the data description part, Unicode encoding is used for processing textual data in any system, and URI, or Uniform Resource Identifier, to univocally identify an abstract or physical resource. Next comes the syntactical base for representing data in a semi-structured fashion, using XML and its associated standard for namespaces and definition of types, XMLSchema. RDF (Resource Description Framework) addresses semantics requirements. It forms a foundation for processing metadata and to express relationships. The Ontology Vocabulary layer uses an ontology language to formally describe the meaning and the terminology used in Web documents. OWL (Web Ontology Language) is likely to become the standard for this layer. The Digital Signature layer gives data a certificate that guarantees their origin. The Logic layer establishes a logical system through which the Proof layer can perform inferences about the data represented in lower layers. Digital Signature combined with Proof assures the validity of the information to be derived in the Trust layer.

The services stack defines distinct service layers. The XML-based Messaging layer provides a message formatting protocol, based upon usual network protocols, offering a high level abstraction for composing and exchanging messages formatted in an XML compliant language. SOAP (Simple Object Access Protocol) is the standard recommended by W3C for this layer. The Service Description layer provides a way to describe Web Services capabilities and communication interfaces. WSDL (Web Service Description Language) is the standard for this layer. Service Publication and Discovery using UDDI (Universal Description, Discovery and Integration) as a standard provide means to make Web Services reachable. The OWL-S language [39] is being proposed as a complement to service description, publication, discovery and composition standards and can even replace them at some degree. Quality of service, security and management are issues that must be considered at every layer of the Services stack.

The Service Flow layer is responsible for coordinating the composition of Web Services in order to achieve a specific functionality. Several standards have been proposed for this layer. They are of special interest to our work and are discussed in Section 2.5.

2.5   Workflow Interchange Standards

Workflows play a major role in constructing applications across the Web, helping to compose and coordinate Services. Currently, there are two main approaches being used to represent workflows on the Web. The first is to directly use an XML-based specification. The other favors functionality, by proposing means of composing services. Since we use workflows for *How-documentation*, and these workflows must support execution, we need to consider how to represent them for a distributed execution on the Web.

There are two major proposals of XML-based languages to represent workflows: XPDL (XML Process Definition Language) [45], and BPEL4WS (Business Process Execution Language for Web Services) [4]. The first was created explicitly to represent workflows in an accessible language. The latter was introduced to meet the requirements of service composition on the Web, using workflow concepts. These two viewpoints generated different, though overlapping, solutions.

XPDL aims at providing a "lingua franca" to represent workflows, enabling different Workflow Management Systems to use the same process specifications. BPEL4WS was introduced as a language to represent service flow coordination and is based on the merge of two other coordination standards, namely IBM's WSFL [22] and Microsoft's XLANG [38]. Recently, BPEL4WS was turned over to a committee of the Oasis-Open consortium [27], which changed its name to WSBPEL (Web Services Business Process Execution Language), and will be responsible for evolving the standard from now on.

Other languages include BPML (Business Process Modeling Language) [3] and the WSCI (Web Service Choreography Interface) [49]. These two languages have different scope than WSBPEL. Whereas BPML has a broader application context, WSCI is restricted to defining roles of services in a composition and needs not understand the whole process definition. Comparison and evaluation of these workflow representation proposals appears in [40, 48]. As discussed in Section 4, we adopt WSBPEL for publishing our *How-documents* on the Web.

3   Specification of Documents for Environmental Planning

Section 2 established the theoretical foundation for our proposal, discussing document management and Semantic Web issues. This Section presents the structures we propose for environmental planning documentation, namely, a hypermedia model to represent *What* documents, scientific workflows to represent *How-documents*, and design rationale structures to store *Why-documents*. The notation used to present the models is based on the entity relationship diagram for simplicity sake. All documents are stored in database tables to be published on the Web. Their integration is supported via additional entities, as well as by links within *What-documents*. For more details the reader is referred to [29, 33]. This Web representation uses XMLSchema (see Section 4).

3.1   Hypermedia Data Model: *What*

A *What* document describes the environmental plan itself – i.e., it supplies a general vision about what was done in the planning activity describing, for example, the plan objectives and methodology used for solution. The choice of a hypermedia model to document *What* data was based on two main factors: (1) it allows organizing documents in a non linear manner, thus facilitating user interaction and semantic links; (2) it supports incorporating multimedia data and thus remote sensing data, essential in environmental planning.

The hypermedia data model designed to document *What* is based on the Dexter Model [17] and some extensions proposed in DHM [26]. The Dexter Model was chosen because it is a reference standard used in many hypermedia systems and it has a well defined set of elements. Figure 2 shows the ER specification of the proposed model. Its main entities are Hyperdocument, Node, Anchor, Endpoint and Link. These entities have the standard semantics of hypermedia documents. Section 6 shows an example of their use within environmental planning.

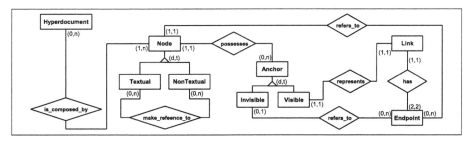

Fig. 2. Hypermedia Data Model for *What* Documentation

A *What* hyperdocument details a problem, and its links point to other (*What*, *Why* or *How*) documents and metadata. Thus, description of a given environmental problem (e.g., whether to allow cutting trees in a preserved area) can be linked to other relevant documented plans (e.g., describing how such an enterprise was successfully conducted in similar conditions).

3.2   Design Rationale Model: *Why*

The *Why* of the decisions in an environmental plan use design rationale. An important aspect in environmental planning is considering the risks presented by some solution alternatives. Risks can be decisive in the choice of an alternative to be implemented. During monitoring/maintenance of an already implemented plan, documentation of risks can be added to explain why a given solution does not work. Thus, in addition to usual design rationale elements, our model supports risk registering for each solution alternative. For instance, in the tree-cutting example, mentioned in Section 3.1, an obvious risk would be the impact on fauna and biodiversity. Figure 3 illustrates the proposed model.

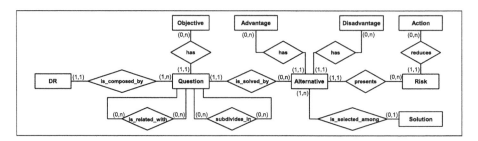

Fig. 3. Design Rationale Model for *Why* Documentation

A *Why-document* is formed by aggregation of questions raised during discussions of a given problem, but that are not necessarily interlinked. A *Question* formalizes a point raised during a design process, for which there are possible *Alternatives*. A *Solution* is the alternative selected for implementation, while an *Objective* is a requirement that should be satisfied by the solution. *Advantage* and *Disadvantage* record positive and negative points concerning an alternative. Any alternative can have *Risks*, and *Actions* can be carried out to reduce a risk.

The meaning of each relationship can be easily comprehended through its name. The relationship <subdivides_in> takes into consideration that complex *Questions* can be solved indirectly by decomposition, i.e., complex questions can be decomposed in more simple questions.

3.3   Scientific Workflow: *How*

The structure designed to represent a scientific workflow to document *How* is an adaptation of the workflow systems standard defined by the Workflow Management Coalition (WFMC). This standard, called Workflow Reference Model, supplies a common generic basis for development of interoperability scenarios between different workflow systems [44].

Figure 4 describes how we record this kind of document. The elements *Workflow, Activity, Atomic Activity, Sub-Workflow, Dependency, Data, Role* and *Software* appear in the WfMC reference model [44], and have the standard meanings. *Data Dependency, Temporal Dependency, Agent* and *User* are new elements introduced in our model to support the needs of environmental applications.

More specifically, a data dependency between two activities is established through exchange of data, with Activity B depending on an Activity A if output data of A constitute input data of B. Temporal dependency determines precedence of execution order of activities in time, e.g. Activity B depends on Activity A if the execution of B cannot start before the ending of A. The notion of sub-workflow allows document reuse – e.g., the plan for determining areas where to cut trees can embed procedures that have been implemented elsewhere.

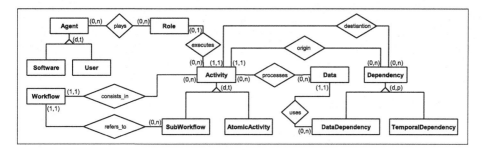

Fig. 4. Specification of a *How-document*

4   Publishing *What-, How-, Why-documents* on the Web

Section 3 shows how we store environmental planning documentation in a database, associating it with metadata. This Section shows how we publish our documents on the Web, enhancing their semantics with ontological associations. Following Semantic Web standards, we adopt XML to represent our documents, and its schema language, XMLSchema, to define their structure and syntactical constraints. This Section presents the schemata in XMLSchema for each of the document types. The specifications presented are partial because of space restrictions. Full schemas can be found in [29]. Section 6 shows examples of XML documents generated.

4.1   Domain Ontologies and Enviromental Documentation

Ontologies are shared elaborated concepts of knowledge about delimited domains [15]. They describe the meaning of terms, or instances, used in a particular domain, according to their defining concepts, or classes, and the semantic relationships among them. Thus, an ontology specifies the possible uses of data and processes, clarifying the usage scope, or context, for the application of these data.

Connecting documents and their components with ontologies improves their significance, especially geographic and environmental planning related ontologies such as introduced in [15, 14]. This connection can be implemented in a simple way by associating documents and URIs of Web available ontologies.

Our approach to combine domain ontologies and environmental documentation is based on the POESIA approach to handle cooperative processes in the Semantic Web [15]. POESIA relies heavily in two concepts: workflows to compose services, and domain ontologies to provide semantics.

The structure of a domain ontology is divided in dimensions that reflect distributed facets. For instance, for the tree cutting example, a spatial dimension defines classes and terms concerning spatial division concepts, a species dimension contains terms that refer to protected species in the area.

Since a term is an instance of an ontology node, terms are unambiguously defined by an ontology path expression, which specifies a unique path in the ontology structure to reach the node. This expression is specified by the concatenated sequence of *concept(term)* vertices visited within the path. As an example, state(Rio).county(Campos) is an unambiguous reference in the spatial dimension to a county called Campos in the state called Rio. An *ontological coverage* is a tuple of unambiguous references to terms of a POESIA ontology. Two examples of ontological coverages are:

(1) [country(Brazil)]
(2) [country(Brazil).state(Rio), species(Leontopithecus rosalia), species(Caiman latirostris)]

In *POESIA*, an ontological coverage determines, for one or more dimensions, the context in which the corresponding data and processes are valid. A term encompasses another if, and only if, it refers to a higher level term within the same dimension. This relation is represented by $\omega \models \sigma$, meaning that ω encompasses σ. Following this reasoning, an ontological coverage Γ encompasses another ontological coverage Δ, or $\Gamma \models \Delta$, if, and only if, for every term $\omega \in \Gamma$ there exists a term $\sigma \in \Delta$ such that $\omega \models \sigma$. In the example of ontological coverages, coverage (1) encompasses (2). Furthermore, (2) refers to endangered species (a kind of monkey and a specific alligator) found in Rio, Brazil, and involves two dimensions: territorial divisions and endangered species.

POESIA's specification of domain ontologies supports clear identification of the concepts involved in environmental planning activities. The notation used to denote relationships and terms is amenable to efficient algorithmic processing in XML database systems. Thus we propose their use in combination with *What-*, *How-* and *Why-documents*, enhancing the semantics of their contents. Encompassing relationships helps reuse – e.g., a solution given to a specific context [country(Brazil)] can be adapted to a context it encompasses [country(Brazil).state(Rio)]. Ontological path expressions can be attached to documents, thereby adding semantics to them.

4.2   *What-Document* Representation

What-documents are specified as hypermedia components, whose Nodes may be distributed on the Web. Their specification for Semantic Web purposes relies on XMLSchema. Domain ontologies can moreover be associated to *What-documents*. A direct mapping from the ER model of Figure 2 can be made to an XMLSchema description. Since this mapping is straightforward because of the similar nature of XML and hypermedia documents, we omit it here.

4.3   *Why-Document* Representation

An XMLSchema specification, partially shown in Table 1, is used for *Why-documents*, mapped from our internal database tables. Ontological references are provided via pointers to an ontological term. More specifically, lines:

"<xsd:element name="ontologyURI" type="xsd:anyURI" minOccurs="0"/>" and

"<xsd:element name="ontologicalCoverage" type="xsd:string" minOccurs="0"/>"

refer to the URI containing the ontology, followed by the corresponding path expression within the ontology, possibly going down to the term level.

Table 1. XMLSchema: *Why-documents* with ontological references

```
<?xml version = "1.0" encoding = "UTF-8"?>

<xsd:schema xmlns:xsd="http://www.w3.org/2001/XMLSchema">

<xsd:element name="dR" type="DRType"/>
<xsd:element name="question" type="QuestionType"/>


<xsd:complexType name="DRType">
    <xsd:sequence>
        <xsd:element name="question" type="QuestionType"
                     minOccurs="1" maxOccurs="unbounded"/>
        <xsd:element name="ontologyURI" type="xsd:anyURI"
                     minOccurs="0"/>
        <xsd:element name="ontologicalCoverage"
                     type="xsd:string" minOccurs="0"/>
        ...
    </xsd:sequence>
    ...
    <xsd:attribute name="dRID" type="xsd:ID"/>
</xsd:complexType>


<xsd:complexType name="QuestionType">
    <xsd:sequence>
        <xsd:element name="drRef" minOccurs="1"/>
        <xsd:element name="isRelatedWithFK" minOccurs="0"
                     maxOccurs="unbounded"/>
        <xsd:element name="subdividesInFK" minOccurs="0"
                     maxOccurs="unbounded"/>
        <xsd:element name="objective" type="ObjectiveType"
                     minOccurs="0" maxOccurs="unbounded"/>
        <xsd:element name="questionString" type="xsd:string"
                     minOccurs="1" maxOccurs="1"/>
        <xsd:element name="ontologyURI" type="xsd:anyURI"
                     minOccurs="0"/>
        <xsd:element name="ontologicalCoverage"
                     type="xsd:string" minOccurs="0"/>
        ...
    </xsd:sequence>
    ...
    <xsd:attribute name="questionID" type="xsd:ID"/>
</xsd:complexType>
...
</xsd:schema>
```

4.4   *How-Document* Representation

How-documents use scientific workflows and link processes, activities and data to ontological coverages. Unlike *What-* and *Why-documents*, they are dynamic - i.e., they can be executed and this execution ensures reuse and adaptation of planning procedures. Thus, it is not enough specify them using XMLSchema. Rather, we must choose a language that allows their execution on the Semantic Web. The problem is that, as mentioned in Section 2.5, there are several standard proposals for workflows on the Web, notably XPDL and WSBPEL. We have chosen the latter because it offers more functionality, and better serves our needs.

This Section presents a brief comparison of these two standards that justifies our choice. For more thorough comparative studies the reader is referred to [40, 41, 47, 48, 36]. No proposal, however, offers all features needed by workflow representation standards, and more work needs to be done in this direction.

XPDL presents several problems [41]. The main issue is that the language lacks support for specifying synchronization constraints. Another issue is what happens when more than one source and/or sink is specified. It is clearly possible to create multiple sources and/or sinks in XPLD, but what is actually executed is not clear. Other features that we need are not supported. Among them we can single out: dynamically determining the number of instances of an activity; specifying choices from outside the document, i. e., from environment variables; the possibility of specifying states; and ways to cancel activities or entire workflows.

Even though WSBPEL has more features that suit our needs, it also presents shortcomings. One of them concerns problems in executing an activity following flow merges. Furthermore, WSBPEL is a loop-blocked language. Within its loop constructs (e. g. while loop), it is not possible to have an arbitrary exit point. This prevents changing the current executing loop block for another. In contrast, XPDL supports non-blocked, loop-blocked and full-blocked classes of workflows, following the definition in [45]. Hence, it is possible to define arbitrary exit points within cycles.

Table 2. WSBPEL extended XMLSchema for *How-documents*

```
<?xml version='1.0' encoding="UTF-8"?>

<schema xmlns="http://www.w3.org/2001/XMLSchema"
        xmlns:wsdl="http://schemas.xmlsoap.org/wsdl/"
                               xmlns:bpws=
  "http://schemas.xmlsoap.org/ws/2003/03/business-process/"
                          targetNamespace=
  "http://schemas.xmlsoap.org/ws/2003/03/business-process/"
                    elementFormDefault="qualified">

<import namespace="http://schemas.xmlsoap.org/wsdl/"
     schemaLocation="http://schemas.xmlsoap.org/wsdl/"/>


<complexType name="tProcess">
    <complexContent>
        <extension base="bpws:tExtensibleElements">
            <sequence>
                <element name="partnerLinks"
                    type="bpws:tPartnerLinks" minOccurs="0"/>
                <element name="partners" type="bpws:tPartners"
                                          minOccurs="0"/>
                <element name="variables"
                    type="bpws:tVariables" minOccurs="0"/>
                <element name="correlationSets"
                    type="bpws:tCorrelationSets" minOccurs="0"/>
                <element name="faultHandlers"
                    type="bpws:tFaultHandlers" minOccurs="0"/>
                <element name="compensationHandler"
                type="bpws:tCompensationHandler" minOccurs="0"/>
                <element name="eventHandlers"
                    type="bpws:tEventHandlers" minOccurs="0"/>
                <xsd:element name="ontologyURI"
                        type="xsd:anyURI" minOccurs="0"/>
                <xsd:element name="ontologicalCoverage"
                        type="xsd:string" minOccurs="0"/>
            </sequence>
            <attribute name="name" type="NCName"
                                   use="required"/>
            <attribute name="targetNamespace" type="anyURI"
                                   use="required"/>
            <attribute name="queryLanguage" type="anyURI"
      default="http://www.w3.org/TR/1999/REC-xpath-19991116"/>
            <attribute name="expressionLanguage" type="anyURI"
      default="http://www.w3.org/TR/1999/REC-xpath-19991116"/>
```

```
            <attribute name="suppressJoinFailure"
                    type="bpws:tBoolean" default="no"/>
            <attribute name="enableInstanceCompensation"
                    type="bpws:tBoolean" default="no"/>
            <attribute name="abstractProcess"
                    type="bpws:tBoolean" default="no"/>
        </extension>
    </complexContent>
</complexType>
...
<complexType name="tInvoke">
    <complexContent>
        <extension base="bpws:tActivity">
            <sequence>
                <element name="correlations"
                    type="bpws:tCorrelationsWithPattern"
                        minOccurs="0" maxOccurs="1"/>
                <element name="catch" type="bpws:tCatch"
                    minOccurs="0" maxOccurs="unbounded"/>
                <element name="catchAll"
                    type="bpws:tActivityOrCompensateContainer"
                                minOccurs="0"/>
                <element name="compensationHandler"
                    type="bpws:tCompensationHandler"
                                minOccurs="0"/>
                <xsd:element name="ontologyURI"
                        type="xsd:anyURI" minOccurs="0"/>
                <xsd:element name="ontologicalCoverage"
                        type="xsd:string" minOccurs="0"/>
            </sequence>
            <attribute name="partnerLink" type="NCName"
                                    use="required"/>
            <attribute name="portType" type="QName"
                                    use="required"/>
            <attribute name="operation" type="NCName"
                                    use="required"/>
            <attribute name="inputVariable" type="NCName"
                                    use="optional"/>
            <attribute name="outputVariable" type="NCName"
                                    use="optional"/>
        </extension>
    </complexContent>
</complexType>
```

Table 2 shows a partial WSBPEL specification within our framework. We assume, for space saving, that the definitions of the corresponding WSDL document are correctly specified.

5   Semantic Web Environmental Planning Support Tool

This Section discusses issues on implementing a system to support documentation of environmental planning activities and their use on the Web. Our proposal is based on the WOODSS system, which supports documentation management and is being ported to the Web.

5.1   The WOODSS System

The documentation ideas presented were implemented for a mono-user environment in WOODSS (WOrkflOw-based spatial Decision Support System) [35, 21], a software developed at the University of Campinas, Brazil. It was developed on top of Idrisi GIS [9] and tested in several environmental planning efforts.

WOODSS is centered on dynamically capturing user interactions with a GIS in real time, and documenting them by means of scientific workflows. It serves three purposes during environmental planning activities: (i) documentation, for reuse and semantics enhancement; (ii) support for decision making; and (iii) construction of a database that describes solutions to planning processes. This paper concerns documentation issues, and therefore only covers the first aspect. Details on other aspects are covered elsewhere [21, 35].

The dynamically generated scientific workflows correspond to the *How-documents* and are stored in a relational database. Users can manipulate, combine and retrieve these workflows, by accessing this database using WOODSS' graphical interface.

At the same time a workflow (the *How-document*) is constructed in WOODSS, planning experts can, at any time, enter data on *What-* and *Why-documents*, by accessing specific system menus. WOODSS also prompts the user for *What-documents* at the beginning and end of a planning session. Finally, users can also specify a *How-document* at a high level by using the graphical interface, without recurring to a GIS. This means that generic procedures can be stored in a database and made available, to be subsequently specialized for specific GIS implementations.

5.2   Extending WOODSS to the Semantic Web

In order to extend WOODSS to work in compliance with the Semantic Web standards we must work on the data and processes discussed in Section 2.4. Section 4 shows our data representation. We use Web Services to construct software modules to allow cooperative environmental planning on the Web using the documents proposed. This Section outlines how to solve these issues.

From the data point of view, our Web compliant data structures are based on XML. Next, semantic relationships among concepts are specified, which can be done in more than one level. The first level uses RDF for data, and RDFSchema for structure and relationship definitions. However a common vocabulary might be needed by some kinds of semantic relationships. OWL (Web Ontology Language) addresses this problem. We replace ontological specification in a language by references to ontologies within documents. More specifically, our solution is to provide additional semantics by links to ontology services, together with term paths. These services know how to interpret these paths. In particular, we use the OntoCover and OntoCarta [14] tools, developed at UNICAMP. OntoCover is a library implemented in Java^{TM} that supports loading, manipulation and visualization of ontologies, making it easy to create references to ontological coverages. OntoCarta is a software being developed to aid navigation on maps, associating context to a spatial dimension ontology. WOODSS coupling to OntoCover associates processes and data with ontological coverages.

Services require the implementation of the layers shown in Section 2.4. Network layer implementations are commonly available. The XML-based messaging layer is supported with SOAP (Simple Object Access Protocol) compliant

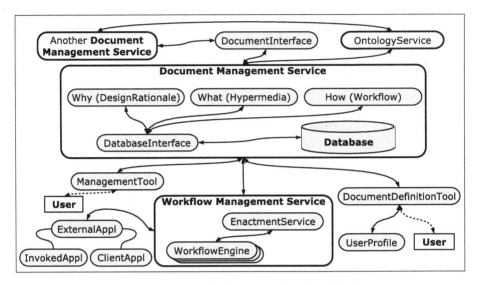

Fig. 5. Document Architecture based on Services

libraries. The service description, publication and discovery layers are provided by standard supporting environment; again, this brings no novelty.

The construction of a Web user interface for WOODSS is another issue to be considered. It involves usability concepts and multiple user management. This discussion is beyond the scope of this work.

Figure 5 shows a high level view of the architecture. It is centered on a Document Management Service that manages document specification and retrieval. The service encapsulates the three kinds of documents, storing them in a relational database. Linkage to ontologies is assured by an interface from the Document Management Service and an Ontology Service that encapsulates the description of a domain ontology. The Document Service also communicates with a Workflow Management Service that extracts the appropriate workflow specifications from it and executes them in workflow engines. Users interact with the Document Service in two ways: via a management tool that supports administrative tasks concerning documents; and via a document definition tool that can be tailored to different user profiles via UserProfile module.

The Workflow Service can invoke external applications (via the ExternalAppl box) and other workflow services. Finally, the Document Management Service can interact with other Document Management Services via a DocumentInterface specification. Each Service blob in the figure can be run at a distinct Web site. Thus, documents can be stored in different databases.

The mono-user version of WOODSS system already supports the functionality of the Document Management Service and its connection to ontologies, workflow management, document definition and management tools. The only external application is the Idrisi GIS. OntoCover and OntoCarta are also implemented and will be encapsulated within an Ontology Service. Thus, the core of our services are already implemented, showing the feasibility of our proposal.

6   Application Example

This Section presents an example of document management within cooperative environmental planning using our proposal.

6.1   Problem Overview

The goal of the problem was to develop an agricultural exploitation plan for a region in Brazil. Planning activities resulted in an agricultural suitability map, showing land suitability in a given region according to a set of relevant parameters. Planners wanted to find areas for agriculture practices within the region, while at the same time taking into account the need for preservation of environmental resources. Input data for solving the problem were maps concerning land use, hydrology, declivity and the result of computing a specific land use model. Figure 6(a) shows the land use map, where the goal region (Iracemápolis microbasin) has areas occupied by pasture, wood and reforestation, water, culture, cities, main and vicinal roads. The capacity of use map was generated by another planning process, previously documented in WOODSS. The declivity map has declivity scales within the area. For these inputs, the problem was solved with support of a GIS and from the solution a *How-document* specification was generated.

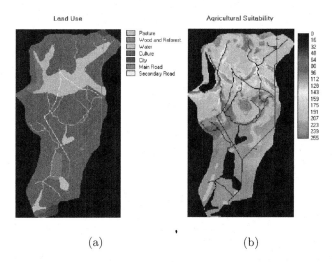

(a)   (b)

Fig. 6. Microbasin of Iracemapolis: (a) Land Use (b) Agricultural Suitability

The result of the planning process is the agricultural suitability map illustrated in Figure 6(b). The best areas for agriculture are those that present the higher values of the scale of values. Areas in solid black cannot be used.

The planning procedure was executed with support of Idrisi GIS. The implementation process consisted on producing several maps with distinct weight

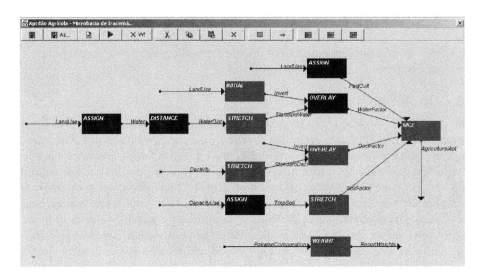

Fig. 7. WOODSS – *Workflow* for *How* Documentation of the example

factors, and overlaying them. Details about functions and the parameters used in each step can be found in [34]. Figure 7 shows the workflow corresponding to the implemented procedure, dynamically generated by WOODSS. In this workflow (executable *How-document*), activities are Idrisi functions and data are files processed by functions. There follows part of the documentation associated with the procedure, along with Semantic Web documents samples.

6.2 *What* Documentation

Figure 8 shows part of the problem's *What* documentation, represented by a hypermedia network of nodes-and-links. This linked structure can be arbitrarily extended to any level of detail (e.g., pointing to formulae and multimedia data). In this example, the main document node (left top corner) describes the general problem. This node is linked to another node that describes the methodology used to solve the problem. This second node contains three visible anchors:

- *next to water*, that points to another node that describes how distance from water was calculated;
- *classification in capacity of use system*, that points to a node describing land classification according to the capacity of use model;
- *lesser declivities*, that points to a node containing a textual description about the procedure used for computing declivity.

In the extended Semantic Web context, each hypermedia node can be in a distinct site, constructed by different users in an asynchronous fashion. Node contents are described in XML, following the XMLSchema specification of Section 4.

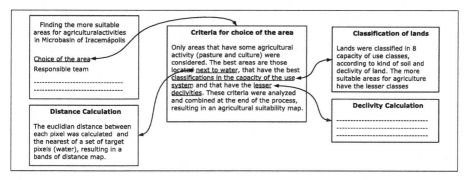

Fig. 8. Partial *What* Documentation for Production of a Solution

6.3   *Why* Documentation

Figure 9 shows part of the *Why* documentation associated with the problem. It describes discussions and decisions related to the choice of restrictions and factors to be considered in solving the decision problem – namely, to find suitable areas for agricultural practices while considering environmental factors. The document is shown as a directed graph. Capital bold letters indicate elements of our design rationale model. (**Q** = question, **SQ** = subquestion, **O** = objective, **A** = alternative, **Ad** = advantage, **D** = disadvantage and **R** = risk). Boxed alternatives were the ones chosen for solution of the corresponding question.

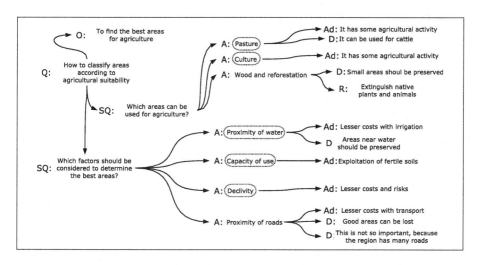

Fig. 9. Partial *Why* Documentation for Problem Solution

Discussion starts with a general question:"How to classify areas according to agricultural suitability?" The objective of this question is *to find the best areas for agriculture*. This complex question is divided into two subquestions:

– Which areas can be used for agricultural practices?
– Which factors should be considered to determine the best areas?

The first subquestion can be answered by three alternatives: Pasture, Culture and Wood; each with advantages, disadvantages and risks. The rest of the figure can be described in the same way.

Table 3. Example of *Why-document* partially translated to XML from Figure 9

```
<?xml version='1.0' encoding="UTF-8"?>

<dR dRID="iracemapolisAreaClassif">
    <question questionID="areaClassif">
        <drRef>iracemapolisAreaClassif</drRef>
        <subdividesInFK>agriUsable</subdividesInFK>
        <subdividesInFK>determFactors</subdividesInFK>
        <objective>To find the best areas for ariculture
        </objective>
        <questionString>How to classify areas according
                            to agricultural suitability?
    </questionString>
    <ontologyURI>
    http://lis.ic.unicamp.br/:8040/ontocover/assess-1034
    <ontologyURI/>
    <ontologicalCoverage>[country(Brazil).state(SaoPaulo),
                            species(Leontopithecus rosalia),
                            species(Caiman latirostris)]
    </ontologicalCoverage>
    </question>
    <question questionID="agriUsable">
    <drRef>iracemapolisAreaClassif</drRef>
    <questionString>Which areas can be used for
                            agriculture?
    </questionString>
    <alternative>
        <altDescription>Pasture</altDescription>
        <advantage>It has some aricultural activity
        </advantage>
        <disadvantage>It can be used for cattle
        </disadvantage>
    </alternative>

<alternative>
    <altDescription>Culture</altDescription>
    <disadvantage>It has some agricultural activity
    </disadvantage>
</alternative>
<alternative>
    <altDescription>Wood and reforestation
    </altDescription>
    <advantage>Short areas that should be preserved
    </advantage>
        <risk>Extinguishiment of native plants and animals
        </risk>
</alternative>
</question>
<question questionID="determFactors">
<drRef>iracemapolisAreaClassif</drRef>
<questionString>Which factors should be considered
                    to determine the best areas?
</questionString>
<alternative>
    <altDescription>Proximity of water</altDescription>
    <advantage>Lesser costs with irrigation</advantage>
    ...
</alternative>
...
</question>
<ontologyURI>
http://lis.ic.unicamp.br/:8040/ontocover/assess-1034
<ontologyURI/>
<ontologicalCoverage>[country(Brazil).state(SaoPaulo),
                    species(Leontopithecus rosalia),
                    species(Caiman latirostris)]
</ontologicalCoverage>
...
</dR>
```

Why-documents are usually centralized, but can be updated by users in different locations. The need for an XML representation of this kind of document goes in the directions of integration, reuse and attaching semantics to the data, which can be embedded in SOAP messages. Table 3 shows the XML representation for our example for the *Why-document*.

6.4 *How* Documentation

Figure 10 shows the *How-document* for the problem, representing the procedure used to solve the problem. This workflow is identical to the one generated by WOODSS (Figure 7); however, components (activities, data and dependencies) are annotated by experts with indications that facilitate the understanding of *How*. Documentation annotation is available in WOODSS.

For example, the longest sequence of activities (third from top to bottom in Figure 10) indicates that: (i) input is the land use map; (ii) the first activity

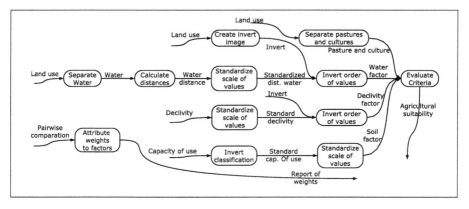

Fig. 10. *How* Documentation for the Problem

(implemented via ASSIGN in Idrisi GIS – see WOODSS workflow, Figure 7) separates water from other elements in the land use map with the water map being passed on to the next step; (iii) the goal of the second activity (DISTANCE in Idrisi GIS) is to compute distance buffers from each point of the region in relation to water; (iv) the third activity (STRETCH in Idrisi GIS) standardizes scales of values of the water distance map, allowing subsequent comparison of all considered factors.

Table 4. Example of *How-document* partially translated to XML from Figure 10

```
<!-- BPEL4WS process definition -->
<process name="agriculturalSuitability"
        targetNamespace="http://lis.ic.unicamp.br/woodss"
                                     xmlns=
    "http://schemas.xmlsoap.org/ws/2003/03/business-process/"
        xmlns:wsdl="http://schemas.xmlsoap.org/wsdl/"
                            abstractProcess="yes">
...
<partnerLinks>
    <partnerLink name="idrisiCaller"
                    partnerLinkType="activityLinkType"
                                myRole="opCaller"/>
    <partnerLink name="idrisiGIS"
                    partnerLinkType="activityLinkType"
                        partnerRole="opResponder"/>
    ...
</partnerLinks>
...
<flow>
    <sequence>
        <invoke  partnerLink="idrisiGIS"
                portType="idrisiCallsPT"
                operation="assignMap"
                inputVariable="inMapPath"
                outputVariable="outMapPath">
                <target linkName="assign-to-gis"/>
                <ontologyURI>
    http://lis.ic.unicamp.br/:8040/ontocover/assess-1034
            <ontologyURI/>
            <ontologicalCoverage>[country(Brazil)]
        </ontologicalCoverage>
            ...
        </invoke>

                    ...
                    <invoke  partnerLink="idrisiGIS"
                            portType="idrisiCallsPT"
                            operation="distanceMap">
                            ...
                    </invoke>
                    ...
                </sequence>
            ...
            </flow>
            <invoke  partnerLink="idrisiGIS"
                    portType="idrisiCallsPT"
                    operation="overlayMaps"
                    inputVariable="inMapPath">
                    ...
            </invoke>
            ...
        </flow>
        <invoke  partnerLink="idrisiGIS"
                portType="idrisiCallsPT"
                operation="evaluateMCE"
                inputVariable="inMapPath">
                ...
        </invoke>
        ...
<ontologyURI>
http://lis.ic.unicamp.br/:8040/ontocover/assess-1034
<ontologyURI/>
<ontologicalCoverage>
[country(Brazil).state(SaoPaulo)]
</ontologicalCoverage>
</process>
```

On the Web context, activities or parts of the workflow can be executed in distinct sites, using various GIS tools. Again, this can be supported by mapping the workflow definition to WSBPEL, as explained in Section 4, and annotating the workflow using XML. Furthermore, each activity and data dependency can refer to an ontology node. Providing this in the Web requires replacing ontology references by [URI, path expression], where the URI points to the ontology server and the path expression to the term within the ontology, as shown in Table 4.

We must point out the essential difference between our proposal and model builder tools such as those provided by ESRI^{TM}[12] software packages. Similar to WOODSS, these packages capture user activities and show them as "workflows" that can be re-executed. However, WOODSS stores these specifications within documentation database tables, thus fostering interoperability and reuse. Therefore, our proposal supports a generic implementation, regardless of the target GIS. First, the document database can be shared and updated by several users simultaneously. Second, the database can store specifications generated for any GIS, since it implements the *Why-document* model of Section 3. A single document database can therefore house models specified within distinct software packages – the only additional requirement is to develop specific modules to encode and decode the commands for each GIS. Our implementation has just one such module – for Idrisi. Extending it to other GIS requires as many additional modules, but data are stored in one database. Finally, generic *How* specifications (such as those of Figure 10) can be defined graphically and stored in the document database, being linked to *What* and *Why* documents, ontologies and metadata. Those generic documents can be exchanged among GIS Web users of any GIS, to be subsequently refined into specific implementations.

7   Concluding Remarks

This paper proposed a framework to support documentation of environmental planning activities in the Semantic Web. It presented three kinds of documents generated during environmental planning: description of the problem to be solved and the associated plan (*What*), description of the process used to obtain the plan (*How*) and description of the reasons behind the planning decisions (*Why*). *What* documents were represented through hypermedia and metadata, *How* documents through scientific workflows and *Why* documents through design rationale.

A mono-user version of this proposal has already been implemented as part of a spatial decision support system – WOODSS – developed at UNICAMP. WOODSS is being extended to meet Semantic Web standards, including connection to ontology terms.

The main contributions are centered on proposing specific document structures for supporting cooperative environmental planning on the Web and an architecture based on Web Services to manage these documents. Documents and processes are linked to each other and associated with geographical metadata and domain ontology terms. Thus, the documents become not only a means of supporting cooperation on the Web, but also of lending more semantics to it.

Another contribution is showing a practical application of the proposal within the Semantic Web context.

Ongoing work involves implementation and theoretical issues. At present, we are implementing the modules responsible for managing the three kinds of documents for the Semantic Web. Issues on the system's user interface should also be considered, given the Web context. This means the interface must support distinct kinds of user profiles and cultures that cooperate on the Web.

Uncertainty is a very important issue in any planning procedure. The present stage of our work does not consider documenting this kind of factor, except via user textual entries in *Why* documentation. Thus, an extension is to provide support to registering probabilities associated with possible outcomes, and use this to help the decision process.

Another extension concerns additional documentation means – e.g., using voice records and video of meetings. These, for example, could be used to generate *Why* documents. Finally, document integration exists at the database level, but must be better reflected at the interface level, to help users query across documents with less navigational effort.

Acknowledgements

This work was developed partially financed by FAPESP, CNPq, the SAI project - Advanced Information Systems - of PRONEX-MCT, as well as WebMaps and AgroFlow CNPq projects. We also thank A. Santanchè, L. Digiampietri and the reviewers for their insightful comments.

References

1. A. Ailamaki, Y. Ioannidis, , and M. Livny. Scientific Workflow Management by Database Management. In *Proc. of 10th IEEE Int. Conf. on Scientific and Statistical Database Management*, pages 190–201, 1998.
2. K. M. Anderson, R. N. Taylor, and E. J. Whitehead Jr. Chimera: hypermedia for heterogeneous software development enviroments. *ACM Transactions on Information Systems (TOIS)*, 18(3):211–245, 2000.
3. A. Arkin. Business Process Modeling Language. Technical report, BPMI.org, 2002. www.bpmi.org (as of Oct 2004).
4. BPEL4WS. Business Process Execution Language for Web Services Version 1.1. Technical report, BEA Systems, International Business Machines Corporation, Microsoft Corporation, SAP AG, Siebel Systems, 2003. www-106.ibm.com/developerworks/webservices/library/ws-bpel/ (as of Oct 2004).
5. T. X. Bui and S. R. Sankaran. Design Considerations for a Virtual Information Center for Humanitarian Assistance/Disaster Relief using Workflow Modeling. *Decision Support Systems*, 31(2):165–179, 2001.
6. J. E. Burge and D. C. Brown. Reasoning with Design Rationale. In *Artificial Intelligence in Design'00*, pages 611–629, 2000.
7. M. A. Casanova, T. A. S. Coelho, M. T. M. Carvalho, E. T. L. Corseuil, H. Nobrega, F. M. Dias, and C. H. Levy. The Design of XPAE - An Emergency Plan Definition Language. In *IV Brazilian Geoinformatics Symp.*, pages 25–32, 2002.

8. F. Casati and A. Discenza. Supporting Workflow Cooperation Within and Across Organizations. In *ACM Symp. Applied Computing*, volume 1, pages 196–202, 2000.
9. Clark Labs. Geographic Analysis and Image Processing Software. www.clarklabs.org (as of Oct 2004).
10. S. Coloudre, T. Libourel, and L. Spéry. Metadata and GIS: a classification of metadata for GIS. In *1st Int. Conf. and Exhibition on Geographic Information*, 1998.
11. J. Conklin and M. Begeman. gIBIS: A Hypertext Tool for exploratory Policy Discussion. *ACM Transactions on Office Information Systems*, 6(4):303–331, 1988.
12. ESRI. GIS and Mapping Service. www.esri.com (as of Oct 2004).
13. Federal Geographic Data Committee. Content Standard for Digital Geospatial Metadata (CSDGM). www.fgdc.gov/metadata/contstan.html (as of Oct 2004).
14. R. Fileto. *The POESIA Approach for Services and Data Integration On the Semantic Web*. PhD thesis, IC–UNICAMP, Campinas–SP, 2003.
15. R. Fileto, L. Liu, C. Pu, E. D. Assad, and C. B. Medeiros. POESIA: An Ontological Workflow Approach for Composing Web Services in Agriculture. *The VLDB Journal*, 12(4):352–367, 2003.
16. G. Fischer, A. Lemke, R. McCall, and A. Morch. *Design Rationale Concepts, Techniques, and Use*, chapter Making Argumentation Serve Design, pages 267–294. Lawrence Erlbaum Associates, 1995.
17. F. Halasz and M. Schwartz. The Dexter Hypertext Reference Model. *Communications of the ACM*, 37(2):30–39, 1994.
18. G. P. Heliades and E. A. Edmonds. Notation and Nature of Task in Comprehending Design Rationale. *Knowledge Based Systems*, 13(4):215–224, 2000.
19. D. L. Hicks, J. J. Legget, P. J. Nürneberg, and J. L. Schnase. A Hypermedia Version Control Framework. *ACM Trans. on Information Systems*, 16(2):127–160, 1998.
20. R. D. Holowczak, S. A. Chun, F. J. Artigas, and V. Atluri. Customized geospatial workflows for e-government services. In *Proc. of the 9th ACM Int. Symp. on Advances in Geographic Information Systems*, pages 64–69, 2001.
21. D. Kaster, C. B. Medeiros, and H. Rocha. Supporting Modeling and Problem Solving from Precedent Experiences: The Role of Workflows and Case-Based Reasoning. *Environmental Modeling and Software*, 2004. Accepted for publication.
22. F. Leymann. Web Services Flow Language (WSFL 1.0). Technical report, IBM, 2001. www-3.ibm.com/software/solutions/webservices/pdf/WSFL.pdf (as of Oct 2004).
23. A. MacLean, R. M. Young, V. Bellotti, and T. Moran. Questions, Options, and Criteria: Elements of design space analysis. *Human-Computer Interaction*, 6(3&4):201–250, 1991.
24. A. MacLean, R. M. Young, and T. P. Moran. Design Rationale: The Argument Behind the Artifact. In *Proc. of the SICCHI Conf. on Wings for the mind*, pages 247–252, 1989.
25. S. R. Monk, I. Sommerville, J. M. Pendaries, and B. Durin. Supporting Design Rationale for System Evolution. In *European Software Engineering Conference (ESEC95)*, pages 307–323, 1995.
26. K. Grønbæk and R. H. Trigg. Design Issues for a Dexter-based Hypermedia System. *Communications of the ACM*, 37(2):40–49, 1994.
27. OASIS Open. OASIS Web Services Business Process Execution Language Technical Committee. http://www.oasis-open.org/committees/tc_home.php?wg_abbrev=wsbpel (as of Oct 2004).

28. J. Oliveira, M. Gonçalves, and C. B. Medeiros. A Framework for Designing and Implementing the User Interface of a Geographic Digital Library. *International Journal of Digital Libraries*, 2(2-3):190–206, 1999.

29. G. Z. Pastorello Jr. Publishing and Integrating Scientific Workflows on the Web (in Portuguese). Master's thesis, IC–UNICAMP, Campinas–SP, 2004.

30. A. Peerbocus, C. B. Medeiros, A. Voisard, and G. Jomier. A System for Change Documentation based on a Spatiotemporal Database. *Geoinformatica*, 8(2):173–204, 2004.

31. C. Petrie and S. Sarin. Internet-based Workflows – Special issue. *Internet Computing*, May-June 2000.

32. I. Rauschert, P. Agrawal, R. Sharma, S. Fuhrmann, I. Brewer, and A. MacEachren. Designing a human-centered, multimodal GIS interface to support emergency management. In *Proc. of the 10th ACM Int. Symp. on Advances in Geographic Information Systems*, pages 119–124, 2002.

33. S. M. Resende. Database-centered Documentation of Environmental Planning Activities (in Portuguese). Master's thesis, IC–UNICAMP, Campinas–SP, 2003.

34. H. A. Rocha. Metadata for Scientific Workflows in Environmental Planning Support (in Portuguese). Master's thesis, IC–UNICAMP, Campinas–SP, 2003.

35. L. Seffino, C. B. Medeiros, J. Rocha, and B. Yi. WOODSS - A Spatial Decision Support System based on Workflows. *Decision Support Systems*, 27(1–2):125–123, 1999.

36. R. Shapiro. A Technical Comparison of XPDL, BPML and BPEL4WS. xml.coverpages.org/Shapiro-XPDL.pdf (as of Oct 2004).

37. M. Singh and M. Vouk. Scientific Workflows: Scientific Workflow Meets Transactional Workflow. In *NSF Workshop on Workflow and Process Automation in Information Systems: State of the Art and Future Directions*, 1996.

38. S. Thatte. XLANG – Web Services for Business Process Design. Technical report, Microsoft, 2001. www.gotdotnet.com/team/xml_wsspecs/xlang-c/default.htm (as of Oct 2004).

39. The OWL Services Coalition. OWL-S 1.0 Release. www.daml.org/services/owl-s/1.0/ (as of Oct 2004).

40. W. M. P. van der Aalst. Don't go with the flow: Web services composition standards exposed. *IEEE Intelligent Systems*, 18(1):72–76, 2003.

41. W. M. P. van der Aalst. Patterns and XPDL: A Critical Evaluation of the XML Process Definition Language. Technical report, Queensland University of Technology, 2003. QUT FIT-TR-2003-06.

42. W3C. The World Wide Web Consortium. www.w3.org (as of Oct 2004).

43. J. Wainer, M. Weske, G. Vossen, and C. B. Medeiros. Scientific Workflow Systems. In *Proc. of the NSF Workshop on Workflow and Process Automation Information Systems*, 1996.

44. WfMC. The Workflow Reference Model. Technical report, Workflow Management Coalition, 1995. TC-1003.

45. WfMC. Workflow Process Definition Interface – XML Process Definition Language. Technical report, Workflow Management Coalition, 2002. TC-1025.

46. E. J. Whitehead Jr. Uniform Comparison of Data Models Using Containment Modeling. In *Proc. of 13th Conf. on Hypertext and Hypermedia*, pages 182–191, 2002.

47. P. Wohed, W. M. P. van der Aalst, M. Dumas, and A. H. M. ter Hofstede. Analysis of Web Services Composition Languages: The Case of BPEL4WS. In *Proc. of the 29th EUROMICRO Conf.*, pages 298–305, 2003.

48. Workflow Patterns. tmitwww.tm.tue.nl/research/patterns/ (as of Oct 2004).
49. WSCI. Web Service Choreography Interface 1.0. Technical report, W3C, BEA Systems, Intalio, SAP AG , Sun Microsystems, 2002. www.w3.org/TR/wsci/ (as of Oct 2004).
50. W. Xin and X. Guangleng. Supporting design reuse based on integrated design rationale. In *IEEE Int. Conf. on Systems, Man, and Cybernetics*, volume 3, pages 1909–1912, 2001.

Semantic Information in Geo-Ontologies: Extraction, Comparison, and Reconciliation

Margarita Kokla and Marinos Kavouras

National Technical University of Athens, 15780 Zografos Campus, Athens, Greece
{mkokla, mkav}@survey.ntua.gr

Abstract. A crucial issue during semantic integration of different geographic metadata sources is category comparison and reconciliation. We focus on the development of a framework for identification and resolution of semantic heterogeneity between geographic categories. The framework is divided in three processes: extraction, comparison and reconciliation. The first process performs semantic information extraction and formalization based on definitions of geographic category terms. Definitions constitute important sources of semantic information for geographic categories. Based on specific rules, definitions are analyzed in a set of semantic elements (properties and values). This information is further used in the second process to identify similarities and heterogeneities between geographic categories. Heterogeneity reconciliation is implemented by semantic factoring, a conceptual analysis process which results in a set of non-redundant, non-overlapping categories.

1 Introduction

Semantic integration constitutes a complicated process in regard to the complex semantics of geographic categories. Categories refer to collections of entities that share common properties. Geographic metadata sources such as categorizations, exchange standards, ontologies, etc., usually include a hierarchy of categories whose meaning and use are described using natural language definitions. A crucial issue during semantic integration of different geographic metadata sources is category comparison and reconciliation. The aim of this process is to identify similarities and resolve heterogeneities between original categories, in order to develop an unambiguous, non-redundant integrated ontology.

Category comparison between two or more ontologies is usually performed with methods or expert-assisting tools, which examine similarity in terms, and hierarchical structure [1, 2, 3]. These methods are sufficient in cases of ontologies with simple semantics, defined mainly by terms and their associated hierarchical relations. In such ontologies, categories are easily perceivable by their terms; consequently similarity in terms expresses similarity in semantics.

Nevertheless, terms and hierarchical relations cannot fully describe the semantics of geographic ontologies. In many cases, linguistic and structural similarity between two categories does not necessarily imply their semantic equivalence. Therefore, category comparison needs to incorporate other available elements of categories that

S. Spaccapietra and E. Zimányi (Eds.): Journal on Data Semantics III, LNCS 3534, pp. 125 – 142, 2005.

may contribute to the accentuation of even slight differences in semantics. Rodriguez and Egenhofer [4] calculate semantic similarity using other features, such as attributes, parts, and functions. This approach is suitable for comparing categories, when both categories have such complete and detailed descriptions. However, most existing geographic metadata sources do not provide this sort of information.

The present research exploits the power of definitions as an alternative superior description of categories' semantics. Definitions constitute an important source of scientific knowledge. They are the primary and usually the only description of geographic categories. Although definitions have been studied a lot in fields such as computational linguistics and lexical semantics, their potential in representing the semantics of geographic information has recently been recognized. Kuhn [5] addresses the issue of explaining the meaning of a term using the notion of conceptual integration from cognitive linguistics. Hakimpour and Timpf [6] present an approach for schema integration based on formal definitions in Description Logic.

The aim of the present work is the development of a general framework for the identification and resolution of semantic heterogeneity. The framework is divided into three processes: extraction, comparison, and reconciliation. The first process exploits methods from the Natural Language Processing (NLP) field in order to extract semantic information from geographic category definitions. It is based on the realization that definitions contain an abundance of semantic information. This can be extracted with appropriate rules to be subsequently formalized in a set of semantic elements. The second process identifies similarities and heterogeneities between geographic categories by comparing terms, semantic elements and their values. Based on these findings, the third process uses a procedure for heterogeneity reconciliation, in order to facilitate the achievement of a truly integrated ontology.

2 Characteristics of Definitions

Previous work [7, 8] has focused on semantic integration of existing geographic metadata sources. Categories were compared using terms, attributes, and hierarchical relations. However, information on attributes and hierarchical relations is not always sufficient for representing the semantics of geographic categories and comparing similar categories. Furthermore, existing sources of geographic information do not include other elements that might contribute to an adequate description of category semantics; functions and parts are examples of such elements. For example, "riverbed", "estuary" and "rapid" are parts of a "river", whereas a "hotel's" function is to "provide rooms and meals for people".

The present work is based on the realization that definitions describe categories' meaning and therefore contain sufficient information to disambiguate similar categories. Geographic information sources use natural language definitions to describe the essential features of their categories. By analyzing geographic category definitions, our aim is to identify and represent in an explicit form the knowledge contained, in order to identify similarities and heterogeneities between similar categories.

Despite some controversy arising from a philosophical debate, definitions are considered essential for the systematization and exchange of general and scientific knowledge. They constitute important sources of knowledge expressed in natural language [9, 10, 11]. Moreover, they retain the meaning of information and reduce vagueness and misinterpretation during information exchange and integration. Research on definitions is seeking ways to exploit the wealth of knowledge immanent in this special kind of text. The NLP field focuses on the automatic extraction of semantic information from definitions [12, 13].

Geographic category definitions contain general and domain knowledge, i.e., knowledge related to the field of geographic information (e.g., land cover, land use, transportation, etc.). As it concerns their generation method, geographic category definitions are intensional, i.e., describe meaning by specifying the essential characteristics or properties of geographic categories, for example:

"well: a hole drilled or dug into the earth or sea bed for the extraction of liquids or gases"

"grassland: area composed of uncultured plants which have little or no woody tissue"

Definitions are considered a kind of text with special structure and content. Intensional definitions are composed of two parts: genus and differentiae. The genus, or hypernym, is the superordinate term of the defined category. For example, in the definition: "river: large natural stream of water", "stream" is the genus of "river".

The differentiae are other elements of the definition apart from the genus, which distinguish terms with the same genus. In the definition: "creek: a small stream, often a shallow or intermittent tributary to a river", "creek" has the same genus as "river", but they are distinguished by the differentiae (e.g., "large", "natural" or "small" and "shallow").

3   Semantic Information Extraction

The methodology adopted for analyzing definitions and extracting immanent semantic information in the form of semantic elements (e.g., LOCATION, PURPOSE, IS-PART-OF, etc.) was introduced by Jensen and Binot [9], and further pursued by Vanderwende [14] and Barriere [13]. This approach is based on the:

1. parsing (syntactic analysis) of definitions, and
2. application of rules that locate certain syntactic and lexical patterns (or defining formulas) in definitions

Parsing determines the structure of a definition, i.e., the form, function, and syntactical relationships of each part of speech. An appropriate tool called parser performs syntactic analysis. The result is usually presented as a parse tree. For the present research, parsing was performed by DIMAP-4 [15], a program for creating and maintaining dictionaries for use in natural language and language technology applications. The program provides functionality for parsing definitions and for identification of basic semantic information, especially IS-A relations. Figure 1 shows the output of the parsing process for the following definition. The symbols used in the parse tree are illustrated in Table 1.

"Body of water: natural or artificial body of water"
```
NP
  adj
      CONJ
          oconj  or
          adj  natural
          adj  artificial
  noun  body
PRP
  prep  of
  NP
  noun  water
```

Fig. 1. Semantic analysis of "Body of water" according to DIMAP-4

Table 1. Symbols used by DIMAP-4

Symbol	Explanation
CONJ	Conjunctive phrase, which may consist of any other type of phrase
NP	Noun Phrase
PRP	Prepositional Phrase
adj	adjective
noun	noun
oconj	ordinary conjunctions: and, but, nor, not, or
prep	Preposition

The parsing result is subsequently used by a set of heuristic rules [16]. These rules examine the existence of syntactic and lexical patterns, i.e., words and phrases in definitions systematically used to express specific semantic information.

In the field of NLP, the notion of "semantic relations" or "thematic roles" is generally used to denote all semantic information extracted generally from text or more specifically from definitions. Here, for conceptual clarity, we further distinguish two types of semantic elements:

1. semantic properties describe characteristics of the category itself (internal characteristics) and
2. semantic relations describe characteristics of the category relative to other categories (external characteristics).

Patterns applied to the genus part of the definition extract the hypernym or IS-A relation. In noun definitions, which are our case, the head of the noun phrase most frequently indicates the genus. However, empty heads, e.g., "kind of", "any of", etc., also indicate the IS-A relation but are not common to geographic category definitions. Patterns applied in the differentiae part extract other semantic information such as: PURPOSE, LOCATION, TIME, SIZE, PART-OF, etc.

Research on automatic acquisition of semantic information has focused more on identification of hypernyms or IS-A relations from definitions [17, 18, 19], free text [20], and the WWW [21] and less on other semantic elements.

Patterns identified in definitions cannot be reliably used for the extraction of the same semantic information from free text. Furthermore, there is no complete list of semantic information that can be extracted from definitions, since they vary according to the dictionary from which they are extracted [13]. Therefore, it was necessary to specify the set of semantic elements used in geographic definitions.

For that reason, different geographic ontologies, standards, and categorizations (e.g., CYC Upper Level Ontology, WordNet, CORINE Land Cover, DIGEST, SDTS,

Table 2. Main semantic prop erties of geographic categories

Semantic Properties
PURPOSE
AGENT
PROPERTY-DEFINED LOCATION
COVER
PROPERTY-DEFINED TIME
POINT IN TIME
DURATION
FREQUENCY
SIZE
SHAPE

Table 3. Main semantic relations of geographic categories

Semantic Relations
IS-A
IS-PART-OF
HAS-PART
RELATIVE POSITION
UPWARD VERTICAL RELATIVE POSITION
DOWNWARD VERTICAL RELATIVE POSITION
IN FRONT OF HORIZONTAL RELATIVE POSITION
BEHIND HORIZONTAL RELATIVE POSITION
BESIDE HORIZONTAL RELATIVE POSITION
SOURCE - DESTINATION
SEPERATION
ADJACENCY
CONNECTIVITY
OVERLAP
INTERSECTION
CONTAINMENT
EXCLUSION
SURROUNDNESS
EXTENSION
PROXIMITY
DIRECTION

etc.) were analyzed in order to identify patterns, which are systematically used to xpress specific semantic elements and formulate the corresponding rules. The most commonly used are shown in Tables 2 and 3. Besides general semantic elements (e.g., PURPOSE, CAUSE, TIME, etc.), other "geographically oriented" elements were also identified. For example, geographic definitions include wealth of information on location (on, below, above, etc.), topology (e.g., adjacent-to, surrounded-by, connected-to, etc.), proximity (e.g., near, far, etc.), and orientation (e.g., north, south, towards, etc). Furthermore, other context-specific semantic elements were also identified. For example, categories relative to hydrography are described by semantic elements such as nature (natural or artificial) and flow (flowing or stagnant). Categories relative to agriculture are described by semantic elements such as crop rotation, tillage, irrigation, vegetation type, etc.

The PURPOSE semantic property is determined by specific phrases containing the preposition "for" (e.g., for (the) purpose(s) of, for, used for, intended for) followed by a noun phrase, present participle, or infinitival clause. The rule for extracting this semantic property from definitions is the following [14]:

If the verb used (intended, etc.) is post-modified by a prepositional phrase with the preposition "for", then there is a PURPOSE semantic property with the head(s) of that prepositional phrase as the value.

For example, the definition "canal: a manmade or improved natural waterway used for transportation" includes a PURPOSE semantic property with "transportation" as the value.

The PROPERTY-DEFINED LOCATION semantic property implies that an action or activity (given by the value of the semantic property) takes place in the defined category. The rule for extracting the semantic property PROPERTY-DEFINED LOCATION is the following [16]:

If the genus term is in the set {place, area, space, ...} and there is a relative clause and the relativizer is in the set {where, in which, on which}, then there is a LOCATION relation between the headword and the verb of the relative clause (along with any of its arguments).

This relation is identified in the following definition: "airfield: a place where planes take off and land".

The HAS-PART semantic relation is determined by phrases such as "consist of", "comprise of", "composed of", and "made of". The rule to extract this semantic relation is formulated as following:

If the verb consist (comprise, compose, etc.) is post-modified by a prepositional phrase with the preposition "of", then there is a HAS-PART semantic relation with the head(s) of that prepositional phrase as the value.

For example, the following definition contains a HAS-PART semantic relation with "road or path" as the value.

"way: artifact consisting of a road or path affording passage from one place to another"

The SURROUNDNESS semantic relation is identified by phrases such as "surrounded by" and "enclosed by" or by phrases containing the prepositions "around", "among" and "between". For example, the following definition includes a SURROUNDNESS semantic relation with "land" as the value.

"lake: body of water surrounded by land"

Based on the above methodology, geographic category definitions are analyzed and formalized according to their semantic elements. Table 4 shows a list of geographic category terms and their definitions related to hydrography derived from the following sources:

- Upper Cyc Ontology: Geography Vocabulary [22]
- WordNet [23]
- The Digital Geographic Information Exchange Standard (DIGEST) [24]: Hydrography

The following definitions have been reformed in order to be compatible with the structure:

"term: genus + differentiae"

Example sentences and phrases enclosed in brackets have been omitted. Due to space limitation, an excerpt of the set of semantic elements and values of the above

Table 4. Geographic category terms and definitions related to hydrography

CYC
Body of Water: natural or artificial body of water
Stream: natural body of fresh water that flows when it is not frozen
River: natural stream of water, normally of a large volume
Lake: land-locked body of water, typically but not necessarily of freshwater
Sea: body of salt water as large as or larger than a lake
Canal: artificial waterway created to be paths for boats, or for irrigation
WordNet
Body of Water: the part of the earth's surface covered with water
Stream: natural body of running water flowing on or under the earth
River: large natural stream of water
Brook: natural stream of water smaller than a river
Lake: body of water surrounded by land
Sea: large body of salt water partially enclosed by land
Way: artifact consisting of a road or path affording passage from one place to another
Watercourse, waterway: conduit through which water flows
Canal: long and narrow strip of water made for boats or for irrigation
Headrace: waterway that feeds water to a mill or water wheel or turbine
DIGEST
Inland Water: inland waterway body
Stream: natural flowing watercourse
Canal: man-made or improved natural waterway used for transportation
Ditch: channel constructed for the purpose of irrigation or drainage
Lake: body of water surrounded by land

Table 5. Example of semantic elements and values for geographic categories

SEMANTIC ELEMENTS

	ORIGINAL CATEGORIES	IS-A	COVER	PURPOSE	NATURE	SURROUNDNESS
CYC	BODY OF WATER	body	water		natural or artificial	
	stream	body	fresh water		natural	
	river	stream	water		natural	
	lake	body	water			land
	sea	body	salt water			
	canal	waterway		boats or irrigation	artificial	
WordNet	BODY OF WATER	part	water			
	stream	body	running water		natural	
	river	stream	water		natural	
	brook	stream	water		natural	
	lake	body	fresh water			land
	sea	body	salt water			land
	WAY	artifact		affording passage …		
	waterway	conduit				
	canal	strip	water	boats or irrigation		
	headrace	waterway		feeds water to a mill …		
DIGEST	INLAND WATER	waterway body				
	stream	watercourse			natural	
	canal	waterway		transportation	man-made or improved natural	
	ditch	channel		irrigation or drainage	constructed	
	lake	body	water			land

Table 6. Value processing

SEMANTIC ELEMENTS

ORIGINAL CATEGORIES		IS-A	COVER	PURPOSE	NATURE	SURROUNDNESS
CYC	BODY OF WATER	body	water		natural or artificial	
	stream	body	fresh water		natural	
	river	stream	water		natural	
	lake	body	water			land
	sea	body	salt water			
	canal	way	water	transportation or irrigation	artificial	
WordNet	BODY OF WATER		water			
	stream	body	running water		natural	
	river	stream	water		natural	
	brook	stream	water		natural	
	lake	body	fresh water			land
	sea	body	salt water			land
	WAY	artifact		affording passage…		
	waterway	way				
	canal	strip	water	transportation or irrigation		
	headrace	way	water	feeds water to a mill…		
DIGEST	INLAND WATER	body	water			
	stream	body	water		natural	
	canal	way	water	transportation	artificial or improved natural	
	ditch	way		irrigation or drainage	artificial	
	lake	body	water			land

geographic categories is shown in Table 5. Thus, each geographic category definition is replaced by a set of semantic elements and their values.

Category comparison is based on terms, semantic elements and corresponding values. However, in order to perform this process, it is necessary to find synonyms and hypernyms for category terms and values. Reference ontologies, dictionaries or thesauri may provide this information, however human intervention may also be necessary at this phase.

For the purpose of our running example, we used WordNet and Merriam-Webster online. For example, values "man-made" and "constructed" are replaced by the synonymous value "artificial". Values "conduit" and "channel" are synonymous and can be replaced by their hypernym "way" (Table 6).

Methodologies for interpreting noun compounds are also valuable at this stage. For example, WordNet defines the term "waterway" as "a navigable body of water". Therefore, this noun compound can be decomposed into two values:

"way" for the IS-A semantic relation and
"water" for the MATERIAL - COVER semantic property.

4 Category Comparison

Category comparison consists in the identification of similarities and heterogeneities between similar categories. This process relies on available elements, which describe categories' semantics, such as terms and definitions. According to the previous section, definitions can be further analyzed into semantic elements and values. Therefore, if we assume that a category definition is analyzed into a set of semantic elements and their corresponding values, then a category C_i is represented by the triple $<T_{Ci}, E_{Ci}, V_{eiCi}>$, where T_{Ci} is the term, E_{Ci} the set of semantic elements and V_{eiCi} the set of corresponding values, i.e.,:

$$E_{C_i} = \left\{ e_{1C_i}, e_{2C_i}, ..., e_{nC_i} \right\}, \tag{1}$$

$$V_{e_iC_i} = \left\{ v_{e_1C_i}, v_{e_2C_i}, ..., v_{e_nC_i} \right\}. \tag{2}$$

Different combinations of T_{Ci}, E_{Ci} and V_{eiCi} lead to four possible comparison cases (expressing degree of equivalence) between two categories:

- equivalence, when the categories are identical in meaning
- difference (non-equivalence), when the categories have different meanings
- subsumption (partial equivalence), when one category has broader meaning than the other
- overlap (inexact equivalence), when categories have similar, but not precisely identical meanings.

In order to cover all possible cases, we assume that terms may be either the same (or synonymous) ($T_{C1} = T_{C2}$) or different ($T_{C1} \neq T_{C2}$). Two sets of semantic elements may be:

Table 7. Comparison cases of T_{Ci}, E_{Ci}, V_{eiCi}

#	Terms	Semantic Elements	Values	Comparison Results	Resolution Action
1	$T_{C1}=T_{C2}$	$E_{C1}=E_{C2}$	$V_{eC1}=V_{eC2}$	equivalence	$C_1=C_2$
2	$T_{C1}=T_{C2}$	$E_{C1}=E_{C2}$	$V_{eC1}\neq V_{eC2}$	homonymy-difference	$C_1\neq C_2$
3	$T_{C1}=T_{C2}$	$E_{C1}=E_{C2}$	$V_{eC1}\otimes V_{eC2}$	overlap	$C_1\otimes C_2$
4	$T_{C1}=T_{C2}$	$E_{C1}=E_{C2}$	$V_{eC1}\supset V_{eC2}$	subsumption	$C_1\supset C_2$
5	$T_{C1}=T_{C2}$	$E_{C1}=E_{C2}$	$V_{eC1}\subset V_{eC2}$	subsumption	$C_1\subset C_2$
6	$T_{C1}=T_{C2}$	$E_{C1}\neq E_{C2}$	…	homonymy-difference	$C_1\neq C_2$
7	$T_{C1}=T_{C2}$	$E_{C1}\subset E_{C2}$	$V_{eC1}=V_{eC2}$	more detailed definition or subsumption	$C_1=C_2$ or $C_1\supset C_2$
8	$T_{C1}=T_{C2}$	$E_{C1}\subset E_{C2}$	$V_{eC1}\neq V_{eC2}$	homonymy- difference	$C_1\neq C_2$
9	$T_{C1}=T_{C2}$	$E_{C1}\subset E_{C2}$	$V_{eC1}\otimes V_{eC2}$	overlap	$C_1\otimes C_2$
10	$T_{C1}=T_{C2}$	$E_{C1}\subset E_{C2}$	$V_{eC1}\supset V_{eC2}$	subsumption	$C_1\supset C_2$
11	$T_{C1}=T_{C2}$	$E_{C1}\subset E_{C2}$	$V_{eC1}\subset V_{eC2}$	subsumption or overlap	$C_1\subset C_2$ or $C_1\otimes C_2$
12	$T_{C1}=T_{C2}$	$E_{C1}\supset E_{C2}$	$V_{eC1}=V_{eC2}$	more detailed definition or subsumption	$C_1=C_2$ or $C_1\subset C_2$
13	$T_{C1}=T_{C2}$	$E_{C1}\supset E_{C2}$	$V_{eC1}\neq V_{eC2}$	homonymy- difference	$C_1\neq C_2$
14	$T_{C1}=T_{C2}$	$E_{C1}\supset E_{C2}$	$V_{eC1}\otimes V_{eC2}$	overlap	$C_1\otimes C_2$
15	$T_{C1}=T_{C2}$	$E_{C1}\supset E_{C2}$	$V_{eC1}\supset V_{eC2}$	subsumption or overlap	$C_1\supset C_2$ or $C_1\otimes C_2$
16	$T_{C1}=T_{C2}$	$E_{C1}\supset E_{C2}$	$V_{eC1}\subset V_{eC2}$	subsumption	$C_1\subset C_2$
17	$T_{C1}=T_{C2}$	$E_{C1}\otimes E_{C2}$	$V_{eC1}=V_{eC2}$	equivalence or overlap	$C_1=C_2$ or $C1\otimes C2$
18	$T_{C1}=T_{C2}$	$E_{C1}\otimes E_{C2}$	$V_{eC1}\neq V_{eC2}$	homonymy- difference	$C_1\neq C_2$
19	$T_{C1}=T_{C2}$	$E_{C1}\otimes E_{C2}$	$V_{eC1}\otimes V_{eC2}$	overlap	$C_1\otimes C_2$
20	$T_{C1}=T_{C2}$	$E_{C1}\otimes E_{C2}$	$V_{eC1}\supset V_{eC2}$	subsumption or overlap	$C_1\supset C_2$ or $C_1\otimes C_2$
21	$T_{C1}=T_{C2}$	$E_{C1}\otimes E_{C2}$	$V_{eC1}\subset V_{eC2}$	subsumption or overlap	$C_1\subset C_2$ or $C_1\otimes C_2$

- equal ($E_{C1}=E_{C2}$),
- different ($E_{C1}\neq E_{C2}$),
- a subset of one another ($E_{C1}\supset\subset E_{C2}$), i.e, one category has more semantic elements than the other,

- overlapping ($E_{C1} \otimes E_{C2}$), i.e., categories have some semantic elements in common. Correspondingly, two sets of values may be:
- equal ($V_{eC1} = V_{eC2}$), when categories have the same values for all semantic elements,
- different ($V_{eC1} \neq V_{eC2}$), when categories have different values for all common semantic elements,
- a subset of one another ($V_{eC1} \supset\subset V_{eC2}$), when the value of at least one semantic element of one category is more general than the other's,
- overlapping ($V_{eC1} \otimes V_{eC2}$), when categories have overlapping values for a semantic element, e.g., V_{eC1} = "irrigation or transportation" and V_{eC2} = "irrigation or drainage". Overlapping sets of values may also occur when $V_{e1C1} \supset V_{e1C2}$ and $V_{e2C1} \subset V_{e2C2}$, i.e., the value of the semantic element e_1 of category C_1 (V_{e1C1}) is more general than the corresponding value of category C_2 (V_{e1C2}), whereas the value of another semantic element e_2 of category C_1 (V_{e2C1}) is more specific than the corresponding value of category C_2 (V_{e2C2}).

Semantic elements and values are not independent. Semantic elements provide the basis for comparing values. Obviously, values are compared only for the same semantic elements. In case of different semantic elements (clear case of different categories), values are not further examined.

Although many combinations between terms, semantic elements and corresponding values may technically occur, in practice comparison is meaningful mainly for semantically similar categories, i.e., categories with the same or synonymous terms and categories with common semantic elements and values.

Table 7 includes indicative, meaningful combinations between T_{Ci}, E_{Ci}, V_{eiCi}, the comparison result and the action required to resolve the case. Examples of some comparison cases between categories of ontologies A and B are given in Table 8. This approach can also prove to be useful in cases where terms are neither equal nor synonymous, but appear to present some similarity in certain semantic elements and their corresponding values. Some of these cases are straightforward and can be easily resolved. For example, categories with $T_{C1} \neq T_{C2}$, $E_{C1} = E_{C2}$ result in:

- an overlap $C_1 \otimes C_2$ when $V_{eC1} \otimes V_{eC2}$,
- a subsumption $C_1 \supset\subset C_2$ when $V_{eC1} \supset\subset V_{eC2}$

Some other cases however, which involve different terms, overlapping sets of semantic elements and overlapping values are complicated and require more detailed analysis and possibly expert's involvement.

Some combinations have two possible comparison results. These require further investigation and expert's involvement in order to discover which the right one for the specific categories is. For example, the combination $T_{C1} = T_{C2}$, $E_{C1} \supset E_{C2}$, $V_{eC1} = V_{eC2}$ can correspond to two cases. In the first case, category C_1 is defined by more semantic elements and therefore is more specific than category C_2. In the second case, the additional semantic elements of C_1 are not primary determinant of its semantics, but rather are context-specific, i.e., relate more to the context and scope of the ontology.

Domain ontologies usually include specialized knowledge, which is not contained in general purpose ontologies. CORINE Land Cover is an example of this case, since definitions describe the way categories are identified from satellite images.

Table 8. Examples of some comparison cases

ONTOLOGY A	ONTOLOGY B
#1: equivalence ($C_1=C_2$)	
T_{C1}: canal	T_{C2}: canal
IS-A: stream	IS-A: stream
PURPOSE: irrigation and transporta-	PURPOSE: irrigation and transportation
NATURE: artificial	NATURE: artificial
#2: difference ($C_1 \neq C_2$)	
T_{C1}: canal	T_{C2}: canal
IS-A: stream	IS-A: path
PURPOSE: irrigation	PURPOSE: transportation
NATURE: artificial	NATURE: natural
#3: overlap ($C_1 \omega C_2$)	
T_{C1}: canal	T_{C2}: canal
IS-A: stream	IS-A: stream
PURPOSE: irrigation and transporta-	PURPOSE: irrigation and drainage
NATURE: artificial	NATURE: artificial
#4: subsumption ($C_1 \supset C_2$)	
T_{C1}: canal	T_{C2}: canal
IS-A: stream	IS-A: stream
PURPOSE: irrigation and transporta-	PURPOSE: irrigation
NATURE: artificial	NATURE: artificial
#7: more detailed definition ($C_1=C_2$) or subsumption ($C_1 \supset C_2$)	
T_{C1}: canal	T_{C2}: canal
IS-A: stream	IS-A: stream
PURPOSE: irrigation and transporta-	PURPOSE: irrigation and transportation
	NATURE: artificial
#10: subsumption ($C_1 \supset C_2$)	
T_{C1}: canal	T_{C2}: canal
IS-A: stream	IS-A: stream
PURPOSE: irrigation and transporta-	PURPOSE: irrigation
	NATURE: artificial
#11: subsumption ($C_1 \subset C_2$) or overlap ($C_1 \omega C_2$)	
T_{C1}: canal	T_{C2}: canal

Table 8. (*continued*)

IS-A: stream	IS-A: stream
PURPOSE: irrigation	PURPOSE: irrigation and transportation
	NATURE: artificial

#19: overlap $(C_1 \varpi C_2)$

T_{C1}: canal	T_{C2}: canal
IS-A: stream	IS-A: stream
PURPOSE: irrigation and transporta-	PURPOSE: irrigation and drainage
NATURE: artificial	AGENT: humans

#22: synonymy $(C_1 = C_2)$

T_{C1}: canal	T_{C2}: channel
IS-A: stream	IS-A: stream
PURPOSE: irrigation and transporta-	PURPOSE: irrigation and transportation
NATURE: artificial	NATURE: artificial

5 Heterogeneity Reconciliation

Table 7 gives an indication of the action required to resolve different comparison cases. These cases are resolved in the reconciliation process, whose purpose is to remove heterogeneities between categories, in order to properly accommodate them in the integrated ontology. Each comparison case is dealt with differently. The first three (equivalence, difference, and subsumption) are easily resolved. In case of equivalence between two categories, a direct correspondence (equality) between them is specified and they appear as one category in the integrated ontology. In the opposite case, i.e., when the categories are different, no correspondence is specified and the integrated ontology includes both categories. In case of a category being more general than another, a subsumption (IS-A) relation is defined in the integrated ontology. The fourth case is the most difficult to resolve. In this case, it is necessary to split the common from the different parts of overlapping categories.

Reconciliation is implemented by a conceptual analysis procedure known as semantic factoring. Semantic factoring decomposes original categories into a set of non-redundant, non-overlapping conceptual building blocks [25]. These building blocks constitute categories themselves and are called semantic factors. The procedure is based on the comparison results of the previous process.

Semantic factoring proceeds bottom-up from specific to general categories. At this point, it is necessary to rely on a general reference ontology, which will provide the most specific categories to initiate the comparison. Each category with no equivalence (partial or exact) is assigned a semantic factor. WordNet's "headrace" is an example of a category, which has neither partially, nor exactly equivalent categories. Then, exactly equivalent categories are assigned a semantic factor. For example, according to

Table 9, category "lake" is equivalently defined among the three original sources (CYC, WordNet and DIGEST). Therefore, these three equivalent categories correspond to one semantic factor, namely g_7 (Table 10).

Table 9. Semantic elements and values of category "lake"

NAME	IS-A	MATERIAL-COVER	SURROUNDED BY
lake (CYC)	body	water	land
lake (WordNet)	body	water	land
lake (DIGEST)	body	water	land

Table 10. Semantic Factoring

	ORIGINAL CATEGORIES	g_1	g_2	g_3	g_4	g_5	g_6	g_7	g_8	g_9	g_{10}
CYC	BODY OF WATER	X	X	X	X			X	X		
	stream	X	X								
	river	X									
	lake							X			
	sea								X		
	canal			X	X						
WordNet	BODY OF WATER	X	X					X	X		
	stream	X	X								
	river	X									
	brook		X								
	lake							X			
	sea								X		
	WAY			X	X		X				X
	waterway			X	X		X				
	canal			X	X						
	headrace						X				
DIGEST	INLAND WATER	X	X	X		X		X		X	
	stream	X	X								
	canal			X		X					
	ditch									X	
	lake							X			

Inexact equivalences are resolved with decomposition of overlapping categories into three semantic factors: one corresponds to the common part and the other two to the different parts. For example, according to Table 11, categories "canal" as defined by CYC, WordNet and DIGEST are not exactly equivalent, but overlap. More specifically, the overlap occurs because of the values of semantic properties PURPOSE and NATURE. The values of these semantic elements will determine the resultant semantic factors. Indeed, three semantic factors (g_3, g_4, and g_5) are defined: g_3 is an artificial transportation canal, g_4 is an artificial irrigation canal and g_5 is an improved natural transportation canal. Therefore, three original overlapping categories are decomposed into three non-overlapping semantic factors.

Table 11. Semantic elements and values of categories "canal"

NAME	IS-A	COVER	PURPOSE	NATURE
canal (CYC)	way	water	transportation or irrigation	artificial
canal (WordNet)	strip (way)	water	transportation or irrigation	
canal (DIGEST)	way	water	transportation	artificial or improved natural

Once semantic factoring of one-factor categories is completed, the process proceeds with more generic categories, which consist of more than one semantic factor. At this phase, in order to properly assign semantic factors to more general categories it is necessary to know intra-ontology relations, i.e., relations between categories of the same ontology resultant from the original hierarchy. For example, WordNet's category "stream" includes two subcategories: "river" and "brook". Therefore, "stream" consists of semantic factors g_1 and g_2, which correspond to those subcategories.

6   Conclusions

Semantic integration relies on the identification and resolution of heterogeneities between categories. However, this process is usually performed empirically based on features that do not completely define category semantics, such as terms and attributes; this may introduce errors in the integrated ontology.

The present work proposes a framework for identifying and resolving heterogeneities between categories of different ontologies, proceeding in three phases: semantic information extraction, category comparison, and heterogeneity reconciliation. The difference from other approaches is that, apart from category terms and hierarchical relations, category semantics immanent in definitions are taken into account for performing the three processes. The methodology is applied to existing geographic ontologies, which are usually defined as hierarchies of category terms with their definitions without additional features like properties, axioms, functions, etc.

Semantic information extraction relies on the position that definitions contain wealth of semantic information expressed in natural language, which can be extracted with specific rules and formalized in a set of semantic elements and values. Category comparison is based on this formalized information and results in one out of four possible comparison cases between categories (equivalence, difference, subsumption, overlap). Reconciliation is performed in order to resolve any heterogeneity between categories in order to result in an unambiguous, non-redundant integrated ontology.

Three are the main advantages of the methodology. First, it formalizes the complicated process of semantic heterogeneity identification and resolution, which is usually performed superficially during ontology integration. Secondly, it takes advantage of definitions, which are a rich source of semantic information for categories. Thirdly, the four possible comparison cases (as degrees of equivalence) describe explicitly the relation between two categories. Especially the case of overlap or inexact equivalence is neither identified, nor resolved in ontology integration approaches.

Geographic ontologies are heterogeneous in many aspects; therefore, in order to maximize explicitness and objectiveness, it is necessary to minimize human intervention during the semantic integration process. Semantic heterogeneity identification neither depends on an expert's subjective decisions, nor on insufficiently defined features. It rather relies on an explicit process based on semantic elements identified with specific rules. However, a process dealing with semantics could not be fully automated. For that reason, future plans include further systematization of the process, specifically as it concerns value processing of semantic elements in order to facilitate their comparison.

References

1. McGuinness, D.L., Fikes, R., Rice, J., Wilder, S.: An environment for merging and testing large ontologies. Proceedings of the Seventh International Conference on Principles of Knowledge Representation and Reasoning (2000)
2. Mitra, P.,Wiederhold, G.: Resolving Terminological Heterogeneity In Ontologies. In Euzenat, J., Gomez-Perez, A., Guarino, N. Stuckenschmidt, H. (eds.): Proceedings of Workshop on Ontologies and Semantic Interoperability at the 15th European Conference on Artificial Intelligence (ECAI) (2002) http://www.cs.vu.nl/~heiner/ECAI-02-WS/Lyon, France
3. Noy, N., Musen, M.: PROMPT: Algorithm and Tool for Automated Ontology Merging and Alignment. Proceedings of the National Conference on Artificial Intelligence (AAAI-2000), Austin, TX, USA (2000)
4. Rodríguez, A., Egenhofer, M.: Determining Semantic Similarity Among Entity Classes from Different Ontologies. IEEE Transactions on Knowledge and Data Engineering 15 (2) (2003) 442-456
5. Kuhn, W.: Modeling the Semantics of Geographic Categories through Conceptual Integration. In Egenhofer, M., Mark, D. (eds.): Proceedings of the 2nd International Conference on Geographic Information Science, GIScience 2002. Lecture Notes in Computer Science, 2478 Springer (2002) 108-118
6. Hakimpour, F., Timpf, S.: A Step towards Geodata Intagration using Formal Ontologies. Proceedings of the 5th AGILE Conference on Geographic Information Science, Palma, Spain (2001)

7. Kokla, M., Kavouras, M.: Fusion of top-level and geographical domain ontologies based on context formation and complementarity. International Journal of Geographical Information Science 15 (7) (2001) 679-687

8. Kavouras, M., Kokla, M.: A method for the formalization and integration of geographical categorizations. International Journal of Geographical Information Science, 16 (5) (2002) 439 – 453

9. Jensen, K., Binot, J.L.: Disambiguating prepositional phrase attachments by using on-line dictionary definitions. Computational Linguistics 13 (3-4) (1987) 251-60

10. Klavans, J., Chodorow, M., Wacholder, N.: Building a Knowledge Base from Parsed Definitions. In Jensen, K., Heidorn, G., Richardson, S. (eds.): Natural Language Processing: The PLNLP Approach , Kluwer Academic Publishers, USA (1993)

11. Swartz, N.: Definitions, Dictionaries, and Meanings (1997)
http://www.sfu.ca/philosophy/swartz/definitions.htm

12. Jensen, K., Heidorn, G., Richardson, S.(eds.): Natural Language Processing: The PLNLP Approach, Kluwer Academic Publishers, USA (1993)

13. Barriere, C.: From a Children's First Dictionary to a Lexical Knowledge Base of Conceptual Graphs. PhD Thesis, School of Computing Science, Simon Fraser University (1997)

14. Vanderwende, L.: The Analysis of Noun Sequences using Semantic Information Extracted from On-Line Dictionaries. Ph.D. thesis, Faculty of the Graduate School of Arts and Sciences, Georgetown University, Washington, D.C. (1995)

15. CL Research: DIMAP-4, Dictionary Maintenance Programs, (2001) http://www.clres.com

16. Dolan, W.B., Vanderwende, L., Richardson, S.D.: Automatically Deriving Structured Knowledge Base from On-line Dictionaries. Proceedings of the Pacific Association for Computational Linguistics, Vancouver, British Columbia, (1993)

17. Chodorow, M.S., Byrd, R.J., Heidorn, G.E.: Extracting semantic hierarchies from a large on-line dictionary. Proceedings of the 23rd Annual ACL Conference, Chicago, Ill. (1985) 299-304

18. Markowitz, J., Ahlswede, T., Evens, M.: Semantically Significant Patterns in Dictionary Definitions. Proceedings of the 24th Annual ACL Conference, New York, N.Y., (1986) 112-119

19. Ide, N., Veronis, J.: Refining Taxonomies Extracted from Machine Readable Dictionaries. In Hockey, S., Ide, N.(eds.): Research in Humanities Computing II,Oxford University Press, (1993) 145-59

20. Grishman, R., Sterling, J.: Acquisition of selectional patterns. Proceedings of COLING-92, Nantes, France (1992) 658-664

21. Sombatsrisomboon, R., Matsuo, Y., Ishizuka, M.: Aquisition of Hypernyms and Hyponyms from the WWW. Proceedings of 2nd Int'l Workshop on Active Mining (AM2003) (in conjunction with Int'l Sympo. on Methodologies for Intelligent Systems), (2003) 7-13,

22. CYCORP, Inc., Upper Cyc Ontology, http://www.cyc.com/

23. WORDNET - a Lexical Database for English, Cognitive Science Laboratory, Princeton University, http://www.cogsci.princeton.edu/~wn/

24. Digital Geographic Information Working Group (DGIWG): The Digital Geographic Information Exchange Standard (DIGEST), Part 4, Annex A: Feature Codes, Edition 2.1 (2000)

25. Sowa, J.: Knowledge Representation: Logical, Philosophical and Computational Foundations Brooks/Cole , USA (2000)

Semantic Mappings in Description Logics for Spatio-temporal Database Schema Integration[⋆]

A. Sotnykova[1], C. Vangenot[1], N. Cullot[2], N. Bennacer[3], and M-A. Aufaure[3]

[1] École Polytechnique Fédérale de Lausanne, Database Laboratory,
CH-1015 Lausanne, Switzerland
{Anastasiya.Sotnykova, Christelle.Vangenot}@epfl.ch
[2] Université de Bourgogne, Laboratoire LE2I, F-21078 Dijon Cedex,
Nadine.Cullot@u-bourgogne.fr
[3] École Supérieure d' Électricité, F-91192 Gif-Sur-Yvette Cedex, France
{Nacera.Bennacer, Marie-Aude.Aufaure}@supelec.fr

Abstract. The interoperability problem arises in heterogeneous systems where different data sources coexist and there is a need for meaningful information sharing. One of the most representative realms of diversity of data representation is the spatio-temporal domain. Spatio-temporal data are most often described according to multiple and greatly diverse perceptions or viewpoints, using different terms and with heterogeneous levels of detail. Reconciling this heterogeneity to build a fully integrated database is known to be a complex and currently unresolved problem, and few formal approaches exist for the integration of spatio-temporal databases. The paper discusses the interoperation issue in the context of conceptual schema integration. Our proposal relies on two well-known formalisms: conceptual models and description logics. The MADS conceptual model with its multiple representation capabilities allows to fully describe semantics of the initial and integrated spatio-temporal schemas. Description logics are used to express the set of inter-schema mappings. Inference mechanisms of description logics allow us to check the compatibility of the semantic mappings and to propose different structural solutions for the integrated schema.

1 Introduction

Information sharing between heterogeneous information sources is a significant challenge, which has been the focus of much research but remains an open problem. Enabling the cooperation of heterogeneous information systems is not easy to achieve because related knowledge is most likely described in different terms and using different assumptions and different data structures. Heterogeneity may

---

[⋆] This work is supported, in the framework of the EPFL Centre for Global Computing, by the Swiss National Funding Agency OFES as part of the European projects KnowledgeWeb (FP6-507482) and DIP (FP6-507483). It is also supported by the MICS NCCR funded by FNRS in Switzerland, under grant number 5005-67322.

S. Spaccapietra and E. Zimányi (Eds.): Journal on Data Semantics III, LNCS 3534, pp. 143–167, 2005.

arise from syntactic, structural and semantic differences in the data sources. *Syntactic heterogeneity* is due to the use of diverse database models (e.g. object oriented vs. relational), *structural heterogeneity* arises from different conceptual choices during the database modeling phase (e.g., modeling as an object, as a relationship, or as an attribute), and *semantic heterogeneity* comes from differences between the terms used to represent information and their intended meaning. Heterogeneity is accentuated for spatio-temporal data, due to the existence of two very different paradigms for data representation, known as the raster mode (space is represented through images) and the vector mode (space is represented as sets of localized objects). Moreover, spatio-temporal data can be represented at different granularities or levels of detail for the spatial and/or temporal features. Also, we have to consider topological relationships between objects, temporal evolution and synchronization relationships.

Two main categories of frameworks have been proposed for the co-operative information systems: *federation of information systems* [1] and *mediation* which relies on the definition of *wrappers* and *mediators* [2]. Wrappers are used to access local sources from the mediation layer and mediators provide a transparent access to the information from the cooperative layer. Mediation-based architectures facilitate evolution through the addition of new data sources. They support cooperation of large information systems and thus they are more suitable in a web environment. Federation-based architectures are best suited for small-scale cooperation. In all these approaches, information sharing can be done either through the definition of direct mappings between the source data sets, or through the definition of an integrated schema together with associated mappings supporting access to the existing data instances.

Irrespectively of the system architecture, a fundamental task in integration is the ability to recognize corresponding information in heterogeneous data sets and to describe the mappings between them. A large number of papers have investigated various facets of mappings, such as mapping discovery, mapping definition or mappings usage.

Mapping Discovery. Surveys originating from two different communities, database and ontology, analyse various propositions from different points of view: database integration [3] and ontology mapping [4]. Work on mapping discovery aims at providing heuristics to find corresponding elements in different information systems, and basically relies on similarity measures. In the survey on automatic schema matching, Rahm and Bernstein in [3] propose a classification of the matching approaches. They distinguish the shema-level and the instance-level matchers. The methods for matching discovery are classified as element-level or structure-level with linguistic or constraint-based heuristics. Automatic mapping discovery became particularly important for ontology cooperation due to the large number of concepts in an ontology. Ehrig and Sure in [5] propose a methodology combining different similarity measures for identifying mappings between two ontologies. Doan et al. in [6] and [7] propose a system, GLUE, that apply machine learning techniques to improve the mapping discovering process. However, complex mappings have proven difficult to extract and the mapping discovery procedure certainly requires human feedback. Dhamankar in [8] presents a promising system, iMAP, for the discovery of complex mappings between database schemas. However, mapping discovery between heterogeneous schemas describing spatio-temporal data still remains an open issue. In the works on ontology mapping described in [4], the sets of mapping operators used in different mapping methods are inferior to the one we use for spatio-temporal schemas mapping; the algorithms used for initial mappings proposition are not designed to capture ontologies spatio-temporal features.

Inter-Schema Correspondences. Complementary to the above approaches, other research works ([9], [10], [11], [12]) focus on formalisms to specify and use inter-schema knowledge. From a conceptual perspective, inter-schema knowledge identifies elements (or sets of elements) in two schemas that describe the same (or related) facts in the real world, and specifies to what extent the data instances and their type definitions relate to each other (i.e., what is identical, what is similar, what is different). This inter-schema knowledge can then be used to build the integrated schema and to provide for an integrated access to the data sources. The four works presented below follow this objective using different languages. A formalism relying on a logic-based language is proposed by Catarci and Lenzerini in [9]. The language they propose is used to describe both schemas and inter-schema knowledge. The reasoning mechanism of the language can then be used to check inter-schema consistency (i.e., the correctness of the cooperative information system) and to support integrated access to data. Calvanese et al. [10] present an architecture for information integration. A Description Logic called $\mathcal{DLR}$, which includes concepts and n-ary relationships, is used to describe the database schemas, to specify inter-schema knowledge; reasoning services are used during the integration process. The same language, $\mathcal{DLR}$, is proposed by Calvanese et al. in [11] to define mappings in a general framework for ontology integration. These mappings allow the mapping of a concept in one ontology to a view, i.e., a query in another ontology. Finally, Devogele et al. [12] propose a complete methodology for spatial database integration based on three phases: schema preparation, correspondence investigation, and integration. The authors also provide an algebraic data manipulation language (algebra for complex objects) to describe inter-schema correspondences that fully supports the description of correspondences between the spatial features of data.

Querying. Once mappings are formally defined, one should be able to use them for query answering and reasoning [13]. Calvanese et al. [11] discuss various approaches for specifying mappings (global- and local-centric approaches) and, for each approach, analyze the complexity of query answering. The authors conclude that mappings should be defined using suitable mechanisms based on query languages. In [14], Halevy et al. express mappings between data sources on a pairwise basis and define inclusion and equivalence relationships between views of each schema. An algorithm enabling queries to go through mappings in order to find data is also proposed.

Semantic Enrichment. In order to reconcile semantic heterogeneity more semantic information about data is needed. Various proposed approaches add extra information to data either through the specification of meta-data, or through the explanation of the context of data or more generally, by using descriptions stored in ontologies. *Meta-data* describe the content of the underlying data in an easily understandable way. *Contexts* are more complex descriptions specifying the domain of source data. *Ontologies*, by definition, provide an encoded representation of a shared understanding of terms and concepts in a given domain and community. They serve as semantic references for users or applications that accept to align their interpretations of the semantics of their data to the interpretation stored in the ontology. Ontologies are actually extensively proposed as a means to overcome interoperability problems [15]. This is the focus of the work of Fonseca et al. in [16]. In their framework, conceptual schemas of geographical databases are mapped to spatial ontologies that are considered as the formal representation of the spatial semantics. The objective in describing such mappings is to enrich the conceptual schema descriptions and thus, to improve the integration of database conceptual schemas. Hakimpour and Geppert in [17] propose a database integration approach that employs formal ontologies merging. Source ontologies (one per database source) are merged by a reasoning system that finds semantic similarity relations between the various definitions used for each concept. An ontology-based schema integrator builds the global schema of the integrated database using the source schemas and the mappings found during the ontology merging process. Fonseca et al. in [18] propose an Ontology-Driven GIS system which plays the role of a system integrator. The idea is to provide access to data by browsing through ontologies. The architecture is based on four main components namely the *ontology server*, the *ontologies*, *mediators* and *applications* that give access to the information sources. The ontology server is the central component providing the connection between the ontologies, the applications and the information sources. The integration is partly realized by the mediators: when the information system is queried, the mediators extract parts of information necessary to generate a complete instance from the ontologies and the information sources.

Our proposal focuses on the co-operation of spatio-temporal databases. In this respect, our objective is to propose a complete methodology for the integration of spatio-temporal conceptual schemas. Our approach relies on two well-known formalisms: conceptual models and description logics. Spatio-temporal conceptual schemas to be integrated are specified using the MADS conceptual data model [19], which can represent rich spatio-temporal semantics. Reasoning services of description logics are then used to check the consistency of the mappings that guide the construction of the integrated schema.

Compared to the papers presented above, our proposal falls within the scope of approaches that aim at defining a formalism or methodology to specify and use inter-schema knowledge. We do not tackle the issue of mapping discovery, as we assume that a set of inter-schema correspondences given by the designer is completed by an inference engine, nor do we consider the subject of query rewriting, which is out of the scope of this work. However, the proposal contributes to the area of research on the following original topics:

- the proposed methodology, based on description logics reasoning mechanisms, conceptual modeling and integrity constraints, is hybrid and thus innovative;
- we are dealing with spatio-temporal data which, to the best of our knowledge, has not yet been attacked;
- we are using reasoning mechanisms of description logics in order to validate the set of inter-schema mappings against the source schemas.

The paper is structured as follows: Section 2 briefly describes the MADS model and introduces two schemas that are used as a running example throughout the paper. Section 3 introduces description logics: The $\mathcal{SHIQ}$ description logic is used to describe source schemas and inter-schema mappings without any spatial and temporal features. The spatial and temporal aspects are specified using an extension of $\mathcal{ALC}$, $\mathcal{ALCRP(D)}$, that provides concrete domain with space and time. Section 4 presents our integration methodology through its various parts: specification of the inter-schema mappings, validation, and generation of an integrated schema. Finally, Section 5 concludes the paper.

2   The MADS Model

MADS [19] is an object+relationship spatio-temporal conceptual data model. In this model, we assume that the real world of interest that is to be represented in the database is composed of complex objects and relationships between them; both characterized by properties (attributes and methods), and both may be involved in a generalization hierarchy (is-a links). To further illustrate our proposal, we will use two MADS schemas shown in Figures 1 and 2. These schemas are designed for two tourist offices describing the same geographical area, the city of Paris. The purpose of the schema T_2 is to provide tourists with information on the closest to the tourist sites boat, bus, metro, and tram stops. Schema T_1 is more general and describes the transport means and the tourist sites of the same city. Both schemas illustrate structural, spatial, and temporal MADS modeling capabilities.

Data structuring capabilities of MADS are orthogonally complemented with space and time modeling concepts, i.e., spatiality and temporality may be associated at the various structural levels: object, attribute, and relationship. The spatiality of an object conveys information about its location and its extent; the temporality describes its lifecycle. For instance in Fig. 1, the object type TouristPlace has both a spatiality (an area) and a temporality (an interval). Attributes may have spatial (e.g. the attribute Start of the object type Walk in Fig. 2) or temporal (e.g. the attribute Season of the object type Theatre) domains of values. A set of predefined spatial and temporal abstract data types is used to describe the spatial and temporal extents of data. The abstract data types are organized in a generalization hierarchy where generic data types are used to describe domains whose values may be of different, more specific types, e.g., small rivers may be described as lines, bigger ones as areas, hence, their domain

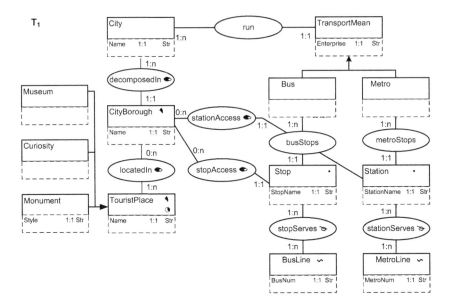

Fig. 1. Schema T_1

should be of the generic type Geo. Attributes may also be space- or time-varying, supporting in this way the continuous view of space. For instance, the attribute nbLanes of Road in Fig. 2 whose value is changing according to the considered road section is a space-varying attribute.

Relationships are either classical n-ary relationships among individual objects or n-ary relationships among sets of objects (multi-association). Relationships may be enhanced with one or several specific semantics, such as aggregation, topological, synchronization, and inter-representation semantics. Topological and synchronization semantics define constraints between spatial and temporal objects respectively. The relationship along between the object types Stop and TransportLine in Fig. 2, holds a topological semantics of intersection.

Multi-representation has been added in MADS as an additional orthogonal dimension. Multi-representation allows the definition in the same schema of several representations for the same real world objects. Those multiple representations may be the consequence of diverging requirements during the database design phase or, in the particular context of spatial data, of the description of data at various levels of detail. The MADS multi-representation feature may also be used in the context of spatial database integration where the full integration, possibly based on different levels of detail, is not possible [12].

To allow users to retrieve specific representations from the set of existing ones, these representations have to be distinguishable and denotable. To this extent, *representation stamps* are added on data, whether they are object type instances or attribute values, and on meta-data, object and relationship type definitions or attribute definitions. Stamps are vectors of values characterizing the context of each representation (e.g. spatial resolution, viewpoint, . . .). Object and relationship types may be representation-varying types and thus have a different set of attributes according to the considered representation. For instance in Fig. 7.d,

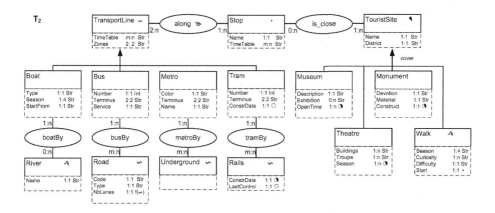

Fig. 2. Schema T_2

the object type Monument is a multi-representation type with two definitions, one for stamp t1 with the attribute Style and one for stamp t2 with the attributes Devotion, Material and Construct. Attributes of such types may have several definitions (different cardinalities and/or value domains) and/or several values (the notation of such an attribute is f(t1, t2) to state that the value is function of the stamp). For instance, the attribute Name of District has a representation-varying definition, i.e., it is a multi-valued attribute for the stamp t1 and a monovalued attribute for the stamp t2. Relationship types may hold several different semantics according to the representation and, for instance, be a topological relationship in one representation and a synchronization in another. We also propose a specific inter-representation semantics that may be applied to both associations and multi-associations to denote that the linked instances are different representations of the same real world object. Actually this inter-representation semantics does not induce any constraints between the linked objects. It denotes paths in the schema that are likely to support consistency checks and update propagation rules. For instance, the correspond relationship in Fig. 7.c holds a multi-representation semantics which states that the instances of TouristSite and TouristPlace linked through this relationship are two representations of the same real world object. For data manipulation, we have defined an algebraic language that provides formal support for manipulating multi-represented data. Concerning multi-representation, users may specify one or several stamps that delimit the subset of the database they will be working on.

3   Description Logics

Description Logics are a family of terminological formalisms with formal logic semantics and designed for representing knowledge and for reasoning about it. Basic elements in a description logic are primitive concepts, primitive roles, the universal concept $\top$ and the bottom concept $\bot$. Complex concepts and roles can be built from primitive ones using the considered description logic constructors. The terminology defines relevant concepts of the domain and their properties. Then individuals occurring in the domain are described using this terminology [20].

Basic description logic constructor, as found in $\mathcal{ALC}$ are: $\neg C$ (negation), $C \sqcap D$ (conjunction), $\forall R.C$ (value restriction) and $\exists R.\top$ (limited existential quantification) where C and D are concepts and R is a role. The $\mathcal{ALC}_{R+}$ description logic is an extension of $\mathcal{ALC}$ with transitive roles. The description logic $\mathcal{SHIQ}$ [21] extends $\mathcal{ALC}_{R+}$ with inverse roles, role hierarchies and qualified number restrictions ($\geq n\ R.C$ and $\leq n\ R.C$). Qualified number restrictions play an important role for representing and for reasoning about conceptual models because they add the ability to model cardinalities of relationships [22]. The expressiveness of $\mathcal{SHIQ}$ is rich and allows encoding of database schemas but it is insufficient to describe spatio-temporal objects.

For representing and reasoning about spatial objects, spatial description logics have been proposed in the literature. Qualitative spatial reasoning in description logic is based on topological relationships [23],[24]. These are known as the set of the $\mathcal{RCC}_8$ relations : Equal (EQ), Disconnect (DC), Externally Connected (EC), Partial Overlap (PO), Tangential Proper Part (TPP), Non-Tangential Proper Part (NTPP) and the inverses of TPP and NTPP : TPPI and NTPPI. A family of description logics called $\mathcal{ALCI}_{\mathcal{RCC}}$ suitable for qualitative spatial reasoning on various granularity is discussed in [25]. The satisfiability problem of these logics is addressed considering the role axioms derived from the $\mathcal{RCC}$ composition tables. Inverse and disjoint roles are also needed to capture the semantics of theses relationships.

Recent work [26] has been proposed to find a way to combine available knowledge representation and reasoning formalisms suitable to consider different real aspects of the world such as time and space. An $\mathcal{E}$-connection is defined in terms of abstract description systems (ADSs), and is a combination of description logics, numerous logics of time and space, and modal and epistemic logics. Link relationships are introduced to combine the formalisms while keeping their domains disjoint. One of the main contributions of the work in [26] is the study on the decidability of the $\mathcal{E}$-connections; it is shown that the $\mathcal{E}$-connections are decidable even with expressive link operators like boolean combinations of link relations.

Extending descriptions logics with concrete domains is a way to introduce new data types such as integer or rational, or to deal with specific dimensions of objects such as spatial or temporal features. The $\mathcal{ALC}(\mathcal{D})$ [27] description logic extends the $\mathcal{ALC}$ DL by adding a new concept-forming predicate operator. $\mathcal{ALC}(\mathcal{D})$ divides the set of objects into two disjoint sets, the abstract and the

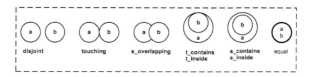

Fig. 3. Topological relationships

concrete objects such as numbers, strings and in particular spatial and temporal objects. Abstract objects can be related to abstract objects via abstract roles and to concrete objects via concrete roles. The relationships between concrete objects are described with a set of domain specific predicates. The pair consisting of a set of concrete objects and the set of predicates forms the concrete domain. Concrete domains increase the expressive power of an extended description logic and allow reasoning on these new features.

The $\mathcal{ALCRP}(\mathcal{D})$ DL proposed by V. Haarslev [28] extends $\mathcal{ALC}(\mathcal{D})$ to build complex roles based on a role-forming predicate operator [29]. In particular, an appropriate concrete domain $\mathcal{S}_2$ is defined for polygons using $\mathcal{RCC}_8$ relations as basic predicates of concrete domain as shown in Fig. 3 (disjoint stands for the DC $\mathcal{RCC}_8$ relationship, touching for EC, s_overlapping for PO, t_inside (t_contains) for TPP, s_inside (s_contains) for NTPP, and equal for EQ). For temporal aspect, the concrete domain $\mathcal{T}$ is a set of time intervals and the 13 Allen relationships (before, after, meets, met-by, overlaps, overlapped-by, during, contains, starts, started-by, finishes, finished-by, equal) are used as basic predicates describing the relationships between intervals. The combination of $\mathcal{S}_2$ and $\mathcal{T}$, $\mathcal{S}_2 \oplus \mathcal{T}$, defines a spatio-temporal concrete domain.

For our purpose, we exploit the $\mathcal{ALCRP}(\mathcal{S}_2 \oplus \mathcal{T})$ expressive power to describe source spatio-temporal schemas and inter-schema mappings. Moreover, the underlying theory allows to detect both inconsistencies and implicit information in the integration process. Using $\mathcal{ALCRP}(\mathcal{S}_2 \oplus \mathcal{T})$, we can define a concept that has a geometry with a specific concrete spatial role called hasArea. Further, using the hasArea feature, we can specify topological relationships between spatial concepts. To define a concept as a temporal concept we can use a specific concrete temporal role called hasDuration. Through this role, temporal relationships between concepts can then be defined. For example, elements of the schema T_1 in Fig. 1 can be described as follows.

A city has a name, it is decomposed in districts and it runs transport means:

City $\sqsubseteq$ ∀Name.String
$\sqcap \geq$ 1Name$\sqcap \leq$ 1Name
$\sqcap$∀decomposedIn.District
$\sqcap$∀run.TransportMean ;

A tourist place has a name, it is a spatio-temporal object thus, it has a geometry and a temporality which are respectively specified by the concrete roles hasArea of the domain Polygon and hasDuration of the domain Interval:

TouristPlace $\sqsubseteq$ $\forall$Name.String
$\sqcap \geq$ 1Name$\sqcap \leq$ 1Name
$\sqcap \exists$hasArea.Polygon
$\sqcap \exists$hasDuration.Interval ;

Museums are tourist places:
Museum $\sqsubseteq$ TouristPlace ;

Monuments are tourist places having a specific feature expressing their style, with the cardinality stating that a monument has exactly one style:

Monument $\sqsubseteq$ TouristPlace
$\sqcap \forall$Style.String
$\sqcap \geq$ 1Style$\sqcap \leq$ 1Style ;

Where, the object types City, Museum, Monument, TouristPlace, District, and TransportMean are modeled as abstract concepts; the relationships decomposedIn, run, and attributes Name, and Style are modeled as roles. Inverse roles can also be defined, for example isRun $\equiv$ run^{-1} . It is also possible to define a contemporary museum as a museum which has at least 10 contemporary paintings:

ContemporaryMuseum $\equiv$ Museum $\sqcap \geq$ 10 expose.ContemporaryPainting ;

To define museums that are spatially connected to some monuments and whose opening times overlap, we first define a spatial predicate connected as the disjunction of elementary predicates, a spatial role spatial_connected based on the previously defined connected predicate, and a temporal role duration_overlaps. The role spatial_connected (respectively duration_overlaps) may be used to link couples of objects whose spatiality (respectively lifecycle) satisfy the connected (respectively overlaps) predicate. Then with these roles, we define such museums, MuseumMonument, as follows:

connected $\equiv$ touching $\vee$ s_overlapping $\vee$ t_contains $\vee$ t_inside $\vee$ s_contains$\vee$
s_inside $\vee$ equal ;
spatial_connected $\equiv$ $\exists$(hasArea)(hasArea).connected ;
duration_overlaps $\equiv$ $\exists$(hasDuration)(hasDuration).overlaps ;

MuseumMonument $\sqsubseteq$ Museum
$\sqcap \exists$spatial_connected.Monument
$\sqcap \exists$duration_overlaps.Monument ;

These descriptions combine not only abstract and concrete objects but also the spatial and temporal concrete domains. This aspect ensures that a GIS system reasoning can be achieved according to the intended semantics of spatiotemporal objects.

4   Integration Methodology

Our integration methodology uses two modeling approaches: database conceptual modeling and modeling in description logics. Fig. 4 shows phases that compose our integration methodology. In the scope of this paper we assume that the source database schemas are expressed in the MADS data model, which has been introduced in Sect. 2. The MADS model has a rich spatio-temporal semantics that is easily understood by a wide circle of designers and users. Contrary to the proposals with rather weak data models enhanced with additional mechanisms for mappings discovery, e.g., as in [8] and [6], we adhere to a different approach, where at the very first phase of the integration procedure, the data are modeled with a very expressive conceptual model. The expressiveness of MADS on one hand greatly simplifies manual mapping discovery, and on the other hand it makes the issue of implementation of mapping discovery algorithms less important in the scope of our integration methodology. Such techniques for MADS model would require very sophisticated algorithms because besides the structural dependencies, i.e., subsumption, there are three more dimensions - spatial, temporal, and multi-representational to be encoded together with their semantics.

MADS was conceived as a conceptual database model, and thus, it lacks reasoning services for the schema integration processes. Defining inter-schema mappings is an error-prone task done manually by the integrated schema designer. Therefore, the compatibility of the set of mappings has to be checked, and to do this task we employ the DL reasoning capabilities. As inter-schema mappings

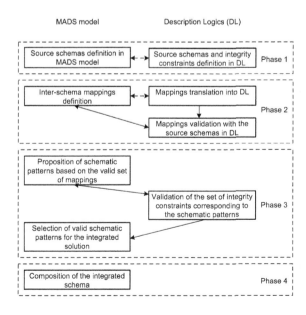

Fig. 4. Integration phases

should be validated against the schemas, the source schemas are translated in description logics (*Phase 1* in Fig. 4). Then the set of inter-schema mappings is translated into description logics (*Phase 2* in Fig. 4).

In our method we differentiate several kinds of inter-schema mappings that are detailed in Sect. 4.1. The set of inter-schema mappings conditions changes that can be potentially applied to the source schemas to construct the final integrated schema. Reasoning services of the DL are used to validate these changes by checking the compatibility of the integrity constraints associated with them (*Phase 3* in Fig. 4).

At the final phase of our method the schema designer is presented with a set of valid schematic patterns that can be used to design the integrated schema (*Phase 4* in Fig. 4). For the running example we present possible structural solutions in Sect. 4.2.

4.1 Inter-schema Mappings

The inter-schema mappings are initially formulated in the MADS language, for details of the language the interested reader may refer to [30]. We distinguish several types of MADS inter-schema mappings. Firstly, there are mappings that express the relationships between populations of schema elements that are intentionally related. We use terms intentionally and extensionally in the same sense as in [9], i.e., intentionally related object types share the schema level representation; extensionally related object types share parts of their populations[4]. We call these mappings *Schema population Correspondences* or SCs. For the population correspondences we apply the set operators shown in Fig. 5. Intuitively, if an SC is asserted, then the intentional equivalence is assumed, and the extensional relationship is defined by the operator. If the operator is disjoint, there are no common instances in the two populations.

Further, another set of mappings describes the intentional relationships between the descriptions of the schema elements involved in an SC. By the description we assume the attributes, including identifiers, and relationships. Since, the attributes and relationships in MADS can have spatial/topological and(or) temporal/synchronization semantics, the set of operators for the set of *Property semantic Correspondences* or PCs includes spatial operators (Fig. 3) and a subset of Allen operators mentioned in Sect. 3. A subset of PCs that involves identifier attributes is called *Matching Rules* or MR. In case of a non-disjoint operator in the SC, the MRs are used to match identical instances. This set of mappings, formulated for all intentionally and extensionally related schema elements is used then for defining possible schematic patterns for an integrated schema.

For the purpose of this paper we will give the inter-schema mappings already translated in DL as presented in Sect. 3. The intention of this translation is to validate the set of inter-schema mappings using inference mechanisms of DL. For this purpose we will weaken some semantic of the mappings, e.g., we use

---

[4] These relations are orthogonal, i.e., two object types can be intentionally equal (the same schema representation) but extensionally disjoint (no common instances) or vice versa.

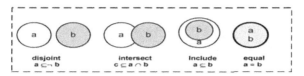

Fig. 5. Population relationships and corresponding DL expressions

the same syntax to state relationships between key and non-key attributes for the validation procedure, but for the clarity of the paper we keep the notion of *Matching Rules* in further discussions.

Schema Population Correspondences. For our running example (schemas T_1 and T_2 in Figures 1 and 2 respectively) the relationship between the population of the object type TouristPlace$_{T_1}$ and TouristSite$_{T_2}$ is not disjoint, as we assume some museums and monuments are represented in both databases. The type of the relationship between these object types is intersection because the subtypes of TouristSite$_{T_2}$, Theatre and Walk, are not modeled in T_1. And, there is a subtype of TouristPlace$_{T_1}$ for which there is no corresponding subtype in TouristSite$_{T_2}$, i.e., the Curiosity subtype. The DL expression stating the intersection of the populations of TouristPlace$_{T_1}$ and TouristSite$_{T_2}$ is SharedTouristSite $\sqsubseteq$ TouristPlace$_{T_1}$ $\sqcap$ TouristSite$_{T_2}$. The populations of Museum$_{T_1}$ and Museum$_{T_2}$, Monument$_{T_1}$ and Monument$_{T_2}$ are included in each other, i.e., in description logics - Museum$_{T_2}$ $\sqsubseteq$ Museum$_{T_1}$, and Monument$_{T_2}$ $\sqsubseteq$ Monument$_{T_1}$.

Already at the level of the *Schema population Correspondences* (SCs), which are the most general correspondences, we can define a set of possible schematic solutions for the integrated schema. Structural patterns that potentially can be applied for constructing the integrated schema, correspond to two decision types that can be taken by the integrated schema designer. The first decision could be to merge overlapping populations. For this decision the inter-schema mappings should be formulated for all related elements and possible structural patterns should be verified. For the second type of decision, where the populations are not merged and therefore, there are no structural transformations to be done, the schema designer is provided with the multi-representation solution. In this case the related populations should be correctly stamped, each instance with the stamp(s) representing the source(s) database(s) they come from; formulated inter-schema mappings become the integrity constraints for the integrated schema. For example, the SC Monument$_{T_2}$ $\sqsubseteq$ Monument$_{T_1}$ would constrain the insert operation by inserting the same instance of Monument in T_1 if an instance of Monument is inserted in T_2.

With the decision required structural changes, the structural pattern cannot be chosen based only upon the relationship between populations, the next essential factor is the *integratability* of the related populations. In other words, the possibility to formulate a valid mapping rule for each related representation in the local schemas. In the set of inter-schema mappings there are two types of correspondences that we call *Property semantic Correspondences* and *Matching Rules* that together with the integrity constraints are meant to assess the integratability of the source schemas. In the next subsection we discuss in more detail *Property semantic Correspondences* and *Matching Rules*.

Property Semantic Correspondences and Matching Rules. With the *Property semantic Correspondences* (PCs), the schema designer states the relationships between different representations (or part of representations) of the intentionally or extensionally same object types. These correspondences are formulated for all the types of the *Schema population Correspondences* (SCs) including disjoint ones, because the PCs relate intentional representations of the object types. The alphabet of the language for the PCs consists in the attribute names of the schema elements involved in the SCs, in other words, the PCs unfold the SCs expressions.

The temporality (lifecycle) of an object is translated in DL by using the predefined role hasDuration and the spatiality by using the role hasArea as defined in [28]. We assume that museums in T_1 have the role hasDuration. In T_2, the temporality is defined through a temporal attribute openTime. Thus in T_2, the museums have a role openTime whose domain is a temporal domain. To express the constraint that says that $openTime_{T_2}$ of $Museum_{T_2}$ is temporally equal to the temporality of $Museum_{T_1}$ we first have to define two roles based on temporal predicates as in [28] :

$$museum\_equal_1 \equiv \exists(openTime_{T_2})(hasDuration).equal ;$$
$$museum\_equal_2 \equiv \exists(hasDuration)(openTime_{T_2}).equal ;$$

Then the constraint is defined as:

$$Museum_{T_2} \sqcap Museum_{T_1} \sqsubseteq \exists museum\_equal_2.Museum_{T_2} \sqcap$$
$$\exists museum\_equal_1.Museum_{T_1} ;$$

To express spatial equality of $TouristPlace_{T_1}$ and $TouristSite_{T_2}$ - both object types have spatial extensions, we state the following expression in DL:

$$area\_equal \equiv \exists(hasArea)(hasArea).equal ;$$
$$TouristPlace_{T_1} \sqcap TouristSite_{T_2} \sqsubseteq$$
$$\exists area\_equal.TouristPlace_{T_1} \sqcap \exists area\_equal.TouristSite_{T_2} ;$$

The rest of the PCs for attributes of $TouristPlace_{T_1}$ and $TouristSite_{T_1}$ are listed below:

$$monument\_equal_1 \equiv \exists(hasDuration)(construct_{T_2}).equal ;$$
$$monument\_equal_2 \equiv \exists(construct_{T_2})(hasDuration).equal ;$$
$$Monument_{T_2} \sqcap Monument_{T_1} \sqsubseteq \exists monument\_equal_2.Monument_{T_2} \sqcap$$
$$\exists monument\_equal_1.Monument_{T_1} ;$$
$$\forall name_{T_1}^{-1}.CityBorough_{T_1} \equiv \forall district_{T_2}^{-1}.TouristSite_{T_2} ;$$
$$\forall name_{T_1}^{-1}.TouristPlace_{T_1} \equiv \forall name_{T_2}^{-1}.TouristSite_{T_2} ;$$

The set of the PCs is complete if for all the SCs stated for the source schemas, all the pairs of elements (attributes) of object and relationship types involved, are examined for the existence of a PC between them. Completeness is ensured by the DL reasoning service. To complete the set of PCs that are initially proposed by the integrated schema designer, an additional set based on the source schema descriptions and the set of the inter-schema mappings, is deduced by the reasoner. Completeness of reasoning means in this context that no valid deduction is left out by the inference engine. In the complete set of the PCs the designer can now state a subset called *Matching Rules* (MRs). The MRs are the rules that state the correspondences between instances that are represented differently in source schemas. These rules involve identifier attributes. Matching rules are useful in order to find corresponding data during the data integration process. For our example, the MRs between $\mathsf{TouristPlace_{T_1}}$ and $\mathsf{TouristSite_{T_2}}$ are those involving $\mathsf{Name_{T_2}}$, $\mathsf{Name_{T_1}}$ attributes and the spatiality of the object types.

$$\mathsf{TouristPlace_{T_1}} \sqcap \mathsf{TouristSite_{T_2}} \sqsubseteq$$
$$\exists\mathsf{area\_equal}.\mathsf{TouristPlace_{T_1}} \sqcap \exists\mathsf{area\_equal}.\mathsf{TouristPlace_{T_2}} ;$$
$$\forall\mathsf{Name_{T_1}^{-1}}.\mathsf{TouristPlace} \equiv \forall\mathsf{Name_{T_2}^{-1}}.\mathsf{TouristSite_{T_2}} ;$$

Validation in DL. As it was mentioned above, we use DL reasoning services to check the satisfiability of our DL model, i.e., the compatibility of the two source schemas, and the set of inter-schema mappings expressed in DL. If our model is found to be unsatisfiable, then the set of the inter-schema mappings should be reconsidered for unsatisfied objects (*Phase 2* in Fig. 4). Unsatisfiability means that there are some concepts that describe an empty set of instances. For our example an unsatisfiable model would be detected for the following set of definitions:

$\mathsf{Stop_{T_1}}$'s are spatially connected to $\mathsf{BusLine_{T_1}}$'s:

$\qquad \mathsf{Stop_{T_1}} \sqsubseteq \exists\mathsf{stopServes_{T_1}}.\mathsf{BusLine_{T_1}} ;$
$\qquad \mathsf{stopServes_{T_1}} \sqsubseteq \mathsf{connected} ;$

$\mathsf{Stop_{T_2}}$'s are spatially connected to $\mathsf{TransportLine_{T_2}}$'s:

$\qquad \mathsf{Stop_{T_2}} \sqsubseteq \exists\mathsf{along_{T_2}}.\mathsf{TransportLine_{T_2}} ;$
$\qquad \mathsf{along_{T_2}} \sqsubseteq \mathsf{connected} ;$

Some stops are represented in both databases described by $\mathsf{T_1}$ and $\mathsf{T_2}$:

$\qquad \mathsf{Stop_{T_1}} \sqsubseteq \mathsf{Stop_{T_2}} ;$

There is no $\mathsf{TransportLine_{T_2}}$ that is spatially connected to a $\mathsf{BusLine_{T_1}}$:

$\qquad \mathsf{area\_disjoint} \equiv \exists(\mathsf{hasArea})(\mathsf{hasArea}).\mathsf{disjoint} ;$
$\qquad \mathsf{TransportLine_{T_2}} \sqsubseteq \exists\mathsf{area\_disjoint}.\mathsf{BusLine_{T_1}} ;$

This model will be invalidated by the reasoner based on the following inferences: firstly, since BusLine and Stop from schema $\mathsf{T_1}$ are spatially connected then, TransportLine and Stop from schema $\mathsf{T_2}$ are also spatially connected. Furthemore, some Stops from $\mathsf{T_1}$ and $\mathsf{T_2}$ are the same, and consequently, some of the $\mathsf{BusLine_{T_1}}$ are spatially connected to $\mathsf{TransportLine_{T_2}}$. This last deduction of the reasoner contradicts the last expression of the model above.

Upon completion of this phase, the schema designer will have in hand a complete and valid set of inter-schema mappings. We are now able to define a set of possible structural solutions for the integrated schema from *Schema population Correspondences*. In the next phase (*Phase 3* in Fig. 4), different schematic patterns will be validated against the compatibility of integrity constraints for the integrated solutions.

4.2   Structural Solution for the Integrated Schema

Proposed schematic patterns for the integrated schema suggest application of a particular structural transformation of the schema elements involved in the inter-schema mappings. These structural transformations should be validated for the integrity of the resulting schema. The question to be answered is, whether these transformations would lead to a violation of the integrity constraints imposed on one or several schemas' elements. If the planned structural transformation is not valid for the given integrity constraints, then the integrity constraints are weakened, or another structural solution is proposed, and the check is run again. To ensure the meaningful integrated solution even for the cases of greatly diverse representations of related data we employ the multi-representation solution consistently preserving the initial representations on the integrated level.

In the following sections we will consider the integration of two object types, $\mathsf{TouristPlace}_{T_1}$, and $\mathsf{TouristSite}_{T_2}$. We assume the following set of correspondences is stated (as explained in Sect. 4.1):

> area_equal $\equiv \exists(\mathsf{hasArea})(\mathsf{hasArea}).\mathsf{equal}$;
> museum_equal$_1 \equiv \exists(\mathsf{openTime}_{T_2})(\mathsf{hasDuration}).\mathsf{equal}$;
> museum_equal$_2 \equiv \exists(\mathsf{hasDuration})(\mathsf{openTime}_{T_2}).\mathsf{equal}$;
> monument_equal$_1 \equiv \exists(\mathsf{hasDuration})(\mathsf{construct}_{T_2}).\mathsf{equal}$;
> monument_equal$_2 \equiv \exists(\mathsf{construct}_{T_2})(\mathsf{hasDuration}).\mathsf{equal}$;

Schema population correspondences:

> (1) $\mathsf{SharedTouristSite} \sqsubseteq \mathsf{TouristPlace}_{T_1} \sqcap \mathsf{TouristSite}_{T_2}$;
> (2) $\mathsf{Museum}_{T_2} \sqsubseteq \mathsf{Museum}_{T_1}$;
> (3) $\mathsf{Monument}_{T_2} \sqsubseteq \mathsf{Monument}_{T_1}$;

Property semantic correspondences:

> (4) $\mathsf{Museum}_{T_2} \sqcap \mathsf{Museum}_{T_1} \sqsubseteq \exists\mathsf{museum\_equal}_2.\mathsf{Museum}_{T_2} \sqcap$
> $\exists\mathsf{museum\_equal}_1.\mathsf{Museum}_{T_1}$;
> (5) $\mathsf{Monument}_{T_2} \sqcap \mathsf{Monument}_{T_1} \sqsubseteq \exists\mathsf{monument\_equal}_2.\mathsf{Monument}_{T_2} \sqcap$
> $\exists\mathsf{monument\_equal}_1.\mathsf{Monument}_{T_1}$;
> (6) $\forall\mathsf{name}_{T_1}^{-1}.\mathsf{CityBorough}_{T_1} \equiv \forall\mathsf{district}_{T_2}^{-1}.\mathsf{TouristSite}_{T_2}$;
>   *Matching Rules* within the *Property semantic correspondences:*
> (7) $\mathsf{TouristPlace}_{T_1} \sqcap \mathsf{TouristSite}_{T_2} \sqsubseteq \exists\mathsf{area\_equal}.\mathsf{TouristPlace}_{T_1} \sqcap$
> $\exists\mathsf{area\_equal}.\mathsf{TouristSite}_{T_2}$;
> (8) $\forall\mathsf{name}_{T_1}^{-1}.\mathsf{TouristPlace}_{T_1} \equiv \forall\mathsf{name}_{T_2}^{-1}.\mathsf{TouristSite}_{T_2}$;

Schematic Patterns. The set of possible schematic patterns depends on the type of the SCs between the related representations. From the spectrum of the structural patterns [31], the integrated schema designer is provided with several patterns for validation. For the context of this paper we chose four structural patterns: fusion - the one resulting in the least number of schema elements for the integrated schema; generalization-partition - the one that produces the most detailed integrated schema; and two types of multi-representations that relate source schemas without changing their structures. For our example schemas, the population correspondence on TouristSite and TouristPlace is intersect (as per assertion (1)), and hence the designer will be provided with all four patterns. The set of available patterns would be different for diferent operators in the population correspondence expression. For example, with the disjoint operator, the generalization-partition pattern is excluded as its application requires common instances in related populations.

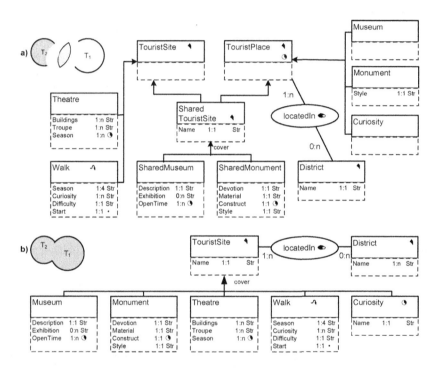

Fig. 6. Schematic solutions under the intersection relation between the populations of the source schemas for integrated schema T_{int}

The first solution (Fig. 6.a) is to extract the overlapping part of the populations and model it as the subtype of the two source populations. This policy uses the multi-inheritance paradigm of the MADS data model. This pattern is called **generalization-partition**. With this structural pattern, the population of the SharedTouristSite$_{T_{int}}$ is TouristPlace$_{T_1}$ $\sqcap$ TouristSite$_{T_2}$. The population of the SharedTouristSite$_{T_{int}}$ are those tourist sites (only of subtypes Museum and Monument) that are close to a public transport stop, i.e., accessible by the public transport.

According to the schema population correspondences (1), (2) and (3), we have an integrated representation for common entities SharedMonument and Shared-Museum for schemas T_1 and T_2. The subtype Curiosity (as well as Theatre and Walk) is not present as a subtype of SharedTouristSite$_{T_{int}}$ because there is no entities of this type neither in Curiosity$_{T_1}$ $\sqcap$ TouristSite$_{T_2}$ nor in the TouristSite$_{T_2}$$\sqcap$ $\neg$Curiosity$_{T_1}$.

From the correspondence assertion stating that the name of the city borough is equal to the district of the TouristSite (assertion (6)), the designer should decide whether he chooses to keep the modeling solution of T_1 - with an object type CityBorough or District, or the modeling solution of T_2 - with an attribute district. In Fig. 6.a, city boroughs are modeled by a spatial object type District with Name attribute. The cardinality of the locatedIn relationship is preserved as it is in schema T_1 - a tourist site can be located in several city boroughs. Such a cardinality would be required for example for the Opera de Paris theatre, that has two buildings, one is in the 9eme city borough, and another in the 12eme. In schema T_2 the cardinality of the District attribute was 1:1, but preserving this cardinality would invalidate the extension (population) of T_1. Another solution for CityBorough would be to keep it as a multivalued attribute of TouristSite (as it is in T_2), but since in T_1 there are relationships attached to the CityBorough object type, the designer should adhere to the pattern shown in Fig. 6.a. where the relationship locatedIn is linked to TouristSite (as in the source schema T_1 for the object type TouristPlace) and attribute District is removed from TouristSite.

Finally, we have to consider the correspondence assertions (4) and (5) about the temporality of TouristPlace and the temporal attributes of Museum and Monument. To be consistent with the schema T_1, the temporality of TouristPlace should be preserved. Considering the MADS model, several solutions are possible for the temporal attributes openTime and construct : we could either remove them as the temporality of TouristPlace will be inherited in Museum and Monument, or define them as derived attributes (derived from the inherited temporality) or finally, keep them and add an integrity constraint. In Fig. 6.a, we choose to present the second possibility with derived attributes to keep the resulting schema more detailed.

The second possible structural pattern (Fig. 6.b) is **fusion** where the populations of the source schemas are merged. As previously, before giving the final schema, the designer has to consider the correspondence assertions (6), (4) and (5). The proposed solution for the CityBorough is the same as in the first pattern for the same reasons. Considering the temporality of the TouristSite object type, the situation and the proposed pattern is different from above: the temporality of TouristPlace$_{T_1}$ is migrated one level down (TouristPlace no longer has a temporality but all its subtypes have one), because in T_2 there are more subtypes for TouristSite and not all of them have temporal attributes. In addition, usage of the redefined temporal attributes OpenTime$_{T_{int}}$ and Construct$_{T_{int}}$ is more expressive for the schema user than would be the inherited temporality (as it was in the source schema T_1).

Finally, the third and fourth possible structural solutions are the **multi-representations** shown in Fig. 7, where the initial representations and local integrity constraints are preserved and no structural transformation is done. This pattern can be applied in the situation where all other proposed patterns are invalidated.

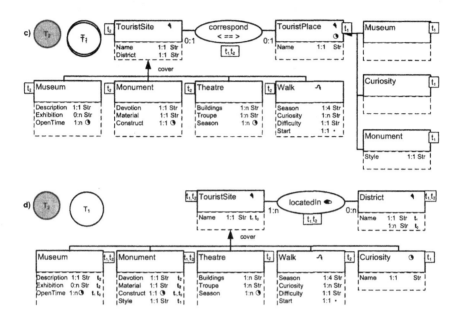

Fig. 7. Multi-representation solutions under the intersection relation between the populations of the source schemas for integrated schema T_{int}

Two possible modeling solutions may be considered: the designer could either choose to link the different object types under consideration with a link holding the specific inter-representation semantics (as in Fig. 7.c) or integrate the different representations in a multi-representation object type (as in Fig. 7.d). The last solution is structurally the same as the fusion but all the schema elements hold the stamps characterizing the schema from where they come: t1 for elements described in T_1, and t2 for elements from T_2. Thus, the object type TouristSite holds the stamps t1, t2 as it is defined in both schemas (with a different name but the same semantic) whereas Curiosity bears only the stamp t1 as it is only described in T1. When considering the object Monument stamped t1, t2, its attributes Devotion and Material are stamped t2 as these attributes are only described in the schema T_2, Style is stamped t1 and finally Construct is defined in both schemas thus stamped t1, t2. Moreover, the object District is stamped t1, t2 and its attribute Name has a representation-varying definition: for t1 it is a monovalued attribute and for t2 it is a multi-valued attribute.

Validation in DL. The compatibility of integrity constraints is checked if the object types under constraints are involved in a SC and according to the chosen structural pattern, the representations of the two concepts are merged (totally or partially). The component ICs must be checked to deduce a common, *global* ICs guaranteeing validity of the resulting *global* ICs. The result of the validation procedure determines if we can define valid integrity constraints for merged object types and consequently for the whole integrated schema.

For our example schemas, assume that a schematic pattern **fusion** proposed for the object types $\mathsf{TransportLine}_{T_2}$ and $\mathsf{BusLine}_{T_1}$ with integrated object type is $\mathsf{TransportLine}_{T_{int}}$ which is defined as $\mathsf{TransportLine}_{T_{int}} \sqsubseteq \mathsf{BusLine}_{T_1} \sqcap \mathsf{TransportLine}_{T_2}$. From the definition of the schema T_1 the reasoner would find that $\mathsf{BusLine}_{T_1}$ has a role $\mathsf{stopServes}_{T_1}$ (Fig. 1) with the cardinalities $\geq 1\mathsf{stopServes}_{T_1}^{-1}.\mathsf{Stop}_{T_1}$ and $\geq 1\mathsf{stopServes}_{T_1}.\mathsf{BusLine}_{T_1}$. On the other side, from the definition of the schema T_2 (Fig. 2) the reasoner would find that $\mathsf{TransportLine}_{T_2}$ has a role along_{T_2} with the cardinalities $\geq 1\mathsf{along}_{T_2}^{-1}.\mathsf{Stop}_{T_2}$ and $\geq 2\mathsf{along}_{T_2}.\mathsf{TransportLine}_{T_2}$. Thus, the definition for the integrated concepts $\mathsf{Stop}_{T_{int}}$ and $\mathsf{TransportLine}_{T_{int}}$ are the following:

$$
\mathsf{TransportLine}_{T_{int}} \sqsubseteq \\
\forall \mathsf{along}_{T_{int}}^{-1}.\mathsf{Stop}_{T_{int}} \\
\sqcap \geq 1\mathsf{along}_{T_{int}}^{-1}.\mathsf{Stop}_{T_{int}} \\
\sqcap \geq 2\mathsf{along}_{T_{int}}^{-1}.\mathsf{Stop}_{T_{int}} ;
$$

$$
\mathsf{Stop}_{T_{int}} \sqsubseteq \\
\forall \mathsf{along}_{T_{int}}.\mathsf{TransportLine}_{T_{int}} \\
\sqcap \geq 1\mathsf{along}_{T_{int}}.\mathsf{TransportLine}_{T_{int}} \\
\sqcap \geq 1\mathsf{along}_{T_{int}}.\mathsf{TransportLine}_{T_{int}} ;
$$

As the resulting integrated cardinality for $\mathsf{along}_{T_{int}}$ the reasoner would propose $\geq 2\mathsf{along}_{T_{int}}^{-1}.\mathsf{Stop}_{T_{int}}$. The choice of this cardinality as the global one may invalidate a part of the population of $\mathsf{BusLine}_{T_1}$ because the cardinality of the $\mathsf{along}_{T_1}^{-1}$ role was 1:n. To meet the cardinality 2:n for all instances of T_{int}, designer could formulate and execute a query that would fill the reference attribute for $\mathsf{TransportLine}_{T_{int}}$ with at least 2 values.

As well, the designer could choose to impose 1:n as the global cardinality of the $\mathsf{along}_{T_{int}}^{-1}.\mathsf{Stop}_{T_{int}}$. In this case, no population is invalidated, but the semantics of 2 terminus stops is lost. To keep this information for the integrated schema, the designer can formulate a query that finds the 2 terminus stops for each instance of the $\mathsf{TransportLine}_{T_{int}}$, so every time the user wants to find two terminus stops for a given line, he/she would execute this query.

4.3   Composing Integrated Schema

By the completion of the validation procedure for the DL integrated schema descriptions, the designer of the integrated schema has valid structural solutions for the related representations assured by the reasoning engine; associated integrity constraints; and mappings for all related elements of the schemas provided by the complete set of inter-schema mappings. In this last phase of the integration process the designer can choose the integrated solutions for each related element of the source schemas and compose the resulting integrated schema.

As it was shown in Sect. 4.2, for each set of mappings, a designer is provided with one or more valid structural solutions (Figures 6 and 7 show possible structural solutions for TouristPlace and TouristSite object types). For the final integrated schema, for each set of mappings for (at least) intentionally related object types, schema designer can choose one of the solutions following a criterion. This criterion is application dependent and could be for example, the complexity of the structural solution, or the type of links used, or the types of queries that will be processed by the information system under development. Considering the structural solutions shown in Figures 6 and 7, the designer could choose the **fusion** as the least complex one, i.e., the one with the least number of elements. As the solution for the bus lines and bus stops representation, the designer could choose to adhere to the **multi-representation** solution as shown

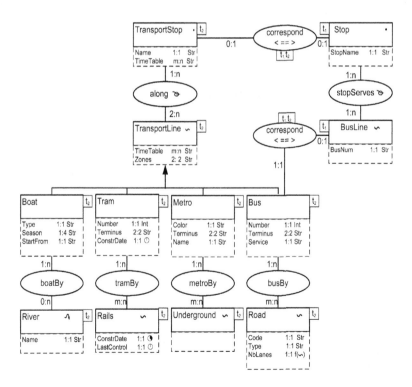

Fig. 8. Multi-representation solution for bus lines and bus stops in an T'_{int}

in Fig. 8, to avoid the invalidation of the population of $BusLine_{T_1}$ object type (cf. the validation section above).

Let us now demonstrate how the mappings can be used in an integrated database. For the part of the schema shown in Fig. 8 the following mappings for the Bus and BusLine object types are relevant:

Schema population correspondences:
(1) $Bus_{T_2} \sqsubseteq BusLine_{T_1}$;
Matching Rules within the *Property semantic correspondences:*
(2) $\forall number^{-1}.Bus_{T_2} \equiv \forall busNum^{-1}.BusLine_{T_1}$;

Let us assume that we use an SQL like query language to query and maintain the database. Mapping (1) requires insertions of an equal instance in the BusLine table every time a new instance is inserted in the Bus table. The equality between the instances of Bus and BusLine is defined by the mapping (2), i.e., the value of the attribute busNum in BusLine must be equal to the value of the attribute number in Bus. Then, the following code will be added to keep the integrated database valid:

```
CREATE TRIGGER busLine
AFTER INSERT ON Bus
   BEGIN
     SELECT busNum FROM BusLine WHERE busNum = NEW.number
       IF SQL%NOTFOUND THEN
       INSERT INTO BusLine VALUES(NEW.number) ;
       ENDIF ;
   END ;
```

Now, every time an insert operation is executed on the Bus table, the trigger will be fired to check the existence of an equal instance in the table BusLine, if it is not found, an insert operation will be executed on the BusLine table, with the value of the busNum attribute equal to the last inserted value of the number attribute.

5   Conclusion

Database integration has been and continues to be the focus of many research efforts, and is a task much harder by the presence of spatio-temporal aspects. Very few formal approaches have been reported which deal specifically with spatio-temporal databases.

In this paper we propose an approach to integrate spatio-temporal database schemas relying on two well-known formalisms: conceptual models and description logics. We use MADS, an object+relationship conceptual model, intended to describe spatio-temporal application data. A peculiar feature of MADS that is of interest in a data integration environment is that it includes specific concepts to describe multiple representations of data. Indeed, as stated in [12], full integration of spatial database requires a powerful data model for the integrated schema in order not to loose the semantics of the original schemas. Description logics are a family of knowledge representation formalisms, with special support for the definition of terminologies. The first phase of our methodology, Figure 4, consists of defining the source schemas using the MADS conceptual model. Inter-schema mappings are then defined between the source schemas. Since defining inter-schema mappings is an error-prone activity, we need to check the compatibility of the mappings. In *Phase 2*, mappings are thus expressed in Description Logics whose inference mechanisms are used for satisfiability checking. From those validated mappings, integration patterns are proposed in *Phase 3*, and their compatibility against integrity constraints is checked. The designer is subsequently provided with a set of valid patterns, to be used in defining the integrated schema (*Phase 4*).

To further this work we plan to design a framework in which one would be able to follow our methodology and to realize its schema integration task in an assisted way. Our framework will combine existing tools like MADS schema editor [19] to design source schemas enhanced with capabilities to define inter-schema mappings; an automatic translator from MADS to a Description logic formalism; and finally, with a DL reasoner like Racer [32] enhanced with spatio-temporal semantics in order to validate the mappings and the integrated schema.

References

1. Sheth, A., Larson, J.: Federated database systems for managing distributed, heterogeneous, and autonomous databases. ACM Computer Surveys (1990) 183–236
2. Garcia-Molina, H., Hammer, J., Ireland, K., Papakonstantinou, Y., Ullman, J., Widom, J.: Integrating and accessing heterogeneous information sources in TSIMMIS. In: Proc. of AAAI Spring Symposium on Information Gathering, Stanford, California (1995)
3. E.Rahm, Bernstein, P.: A survey of approaches to automatic schema matching. The VLDB Journal (2001) 334–350
4. Kalfoglou, Y., Schorlemmer, M.: Ontology mapping: the state of the art. The Knowledge Engineering Review **18** (2003) 1–31
5. Ehrig, M., Sure, Y.: Ontology mapping - an integrated approach. In: Proc. of the ESWS 2004, Heraklion, Crete, Greece (2004) 76–91
6. Doan, A., Madhavan, J., Domingos, P., Halevy, A.: Ontology matching: A machine learning approach. In Staab, S., Studer, R., eds.: Handbook on Ontologies in Information Systems. Springer-Velag (2004) 397–416
7. Doan, A., Madhavan, J., Domingos, P., Halevy, A.: Learning to map between ontologies on the Semantic Web. In: Proc. of International WWW conference, WWW'02, Honolulu, Hawaii, USA (2002)
8. Dhamankar, R., Lee, Y., Doan, A., Halevy, A., Domingos, P.: iMAP: Discovering Complex Semantic Matches between Database Schemas. In: Proc. of the 2004 ACM SIGMOD International Conference on Management of Data, Paris, France, ACM Press (2004) 383–394

9. Catarci, T., Lenzerini, M.: Representing and using interschema knowledge in co-operative information systems. Journal of Intelligent and Cooperative Information Systems **2** (1993) 375–398

10. Calvanese, D., de Giacomo, G., Lenzerini, M., Nardi, D., Rosati, R.: Information integration: Conceptual modeling and reasoning support. In: CoopIS 1998. (1998) 280–291

11. Calvanese, D., de Giacomo, G., Lenzerini, M.: A Framework for Ontology Integration. In: Proceedings of SWWS'01, The First Semantic Web Working Symposium Stanford University, California, USA (2001) 303–306

12. Devogele, T., Parent, C., Spaccapietra, S.: On spatial database integration. International Journal of Geographic Information Systems, Special Issue on System Integration **3** (1998) 335–352

13. Madhavan, J., Bernstein, P., Domingos, P., Halevy, A.: Representing and reasoning about mappings between domain models. In: Proc. of the AAAI Eighteenth National Conference on Artificial Intelligence, Edmonton, Alberta, Canada, American Association for Artificial Intelligence (2002) 80–86

14. Halevy, A.Y., Ives, Z.G., Mork, P., Tatarinov, I.: Piazza: Data Management Infrastructure for Semantic Web Applications. In: Proc. of International WWW Conference, WWW'03. (2003) 556–567

15. Wache, H., Vögele, T., Visser, U., Stuckenschmidt, H., Schuster, G., Neumann, H., Hübner, S.: Ontology-Based Integration of Information - A Survey of Existing Approaches. In: Proceedings of the IJCAI-01 Workshop on Ontologies and Information Sharing, Seattle, USA (2001)

16. Fonseca, F., Davis, C., Câmara, C.: Bridging ontologies and conceptual schema in geographical information integration. Geoinformatica **7** (2003) 355–378

17. Hakimpour, F., Geppert, A.: Global schema generation using formal ontologies. In: Proc. of the ER2002. LNCS 2503 (2002) 307–321

18. Fonseca, F., Egenhofer, M., Agouri, P., Câmara, C.: Using ontologies for integrated geographic information systems. Transactions in GIS **6** (2002) 231–257

19. MurMur project: MurMur consortium: Multi-representations and Multiple resolution in geographic databases. http://lbdwww.epfl.ch/e/MurMur (2002)

20. Baader, F., Calvanese, D., McGuinness, D., Nardi, D., Patel-Schneider, P., eds.: The Description Logic Handbook: Theory, Implementation, and Applications. Cambridge University Press (2003)

21. Horrocks, I., Sattler, U., Tobies, S.: Practical reasoning for expressive description logics. Logic Journal of the IGPL **8** (2000) 161–180

22. Berardi, D., Calvanese, D., de Giacomo, G.: Reasoning on UML class diagrams using description logic based systems. In: Proc. of the KI'2001 Workshop on Applications of Description Logics. (2001) CEUR Electronic Workshop Proceedings, http://ceur-ws.org/Vol-44/.

23. Cohn, A.G., Hazarika, S.M.: Qualitative spatial representation and reasoning: An Overview. Fundamenta Informaticae **46** (2001) 1–29

24. Gerevini, A., Nebel, B.: Qualitative spatio-temporal reasoning with RCC-8 and Allen's interval calculus: Computational complexity. In: Proceedings of ECAI'2002, IOS Press (2002) 312–316

25. Wessel, M.: On spatial reasoning with description logics - position paper. In: Proc. of the International Workshop in Description Logics 2002 (DL2002), Touluse, France (2002)

26. Kutz, O., Lutz, C., Wolter, F., Zakharyaschev, M.: $\mathcal{E}$-connections of abstract description systems. Artif. Intell. **156** (2004) 1–73

27. Baader, F., Hanschke, P.: A scheme for integrating concrete domains into concept languages. In: Proc. of the Twelfth International Joint Conference on Artificial Intelligence (IJCAI), Sydney, Australia (1991) 452–457
28. Haarslev, V., Lutz, C., Möller, R.: A description logic with concrete domains and a role-forming predicate operator. Journal of Logic and Computation **9** (1999)
29. Lutz, C.: Description logics with concrete domains—a survey. In: Advances in Modal Logics Volume 4, King's College Publications (2003)
30. Sotnykova, A., Monties, S., Spaccapietra, S.: Semantic integration in MADS conceptual model. In Bestougeff, H., Thuraisingham, B., eds.: Heterogeneous Information Exchange and Organizational Hubs. Kluwer (2002)
31. Dupont, Y.: Resolving Fragmentation Conflicts in Schema Integration. In Loucopoulos, P., ed.: 13th International Conference on the Entity-Relationship Approach, ER'94. Volume 881 of LNCS., Manchester, U.K., Springer (1994) 513–532
32. Haarslev, V., Möller, R.: Racer system description. In: Proc. of International Joint Conference on Automated Reasoning, IJCAR 2001, Springer-Verlag (2001) 701–705

Data Semantics in Location-Based Services

Nectaria Tryfona and Dieter Pfoser

Research Academic Computer Technology Institute,
Akteou 11, 1851 Athens, Greece
{tryfona, pfoser}@cti.gr, htp://dke.cti.gr

Abstract. As location-based applications become part of our everyday life, ranging from traffic prediction systems to services over mobile phones providing us with information about our surroundings, the call for more semantics and accurate services is emerging. In this work, we analyze and register the data semantics of Location-based Services (LBS). Initially, we categorize LBS data according to the related concepts and use. We distinguish the (a) Domain Data, including spatial and temporal concepts, namely, position, location, movement and time, (b) Content Data, describing the LBS specific content, and (c) Application Data, consisting of the user profile and the services provided by LBS. Next, we model these three data categories in a way that captures their peculiarities and allows their sharing and exchange among different LBS, when desired. For this, we use semantically rich and expressive models, like UML, as well as the long-praised method of ontologies, realized in the open source, ontology and knowledge-based editor Protégé. To argue about the design choices and show their applicability, we present examples from two characteristic real-world applications, both in the Athens Metropolitan Area: an LBS for tourists carrying mobile devices, and a traffic LBS informing drivers about troublesome situations.

1 Introduction

In the recent years Location-Based Services (LBS) enjoy much attention from both the scientific community and the industry. Work has mostly been concentrated on *delivering* information to the mobile user that is related to his/her location and therefore, presumably, more relevant. Additionally, the technological revolution in this area (e.g., advanced capabilities of handheld devices) as well as commercially oriented solutions to customer's needs (e.g., fast transmission of multimedia data such as, images and video) have driven the focus of LBS away from what they really are: services supported by non-conventional databases, characterized by the spatial and temporal dimension, i.e., spatiotemporal databases. Due to this, data involved in LBS have not been really examined in depth. Consequently, LBS data semantics are not captured properly, LBS data models do not fully accommodate application requirements, and the final system does not always meet user needs.

In this work, we treat LBS as non-conventional applications. In these, it is important to understand and register the related concepts. Initially, we analyze the data

S. Spaccapietra and E. Zimányi (Eds.): Journal on Data Semantics III, LNCS 3534, pp. 168–195, 2005.

scenario in LBS and categorize data according to its semantics and use. We distinguish the (a) *Domain Data*, including spatial and temporal concepts, (b) *Content Data*, describing LBS specific content, and (c) *Application Data*, consisting of the user profile and the services provided by the LBS. The goal is to model these three data categories in a way to (i) capture their peculiarities and (ii) allow the sharing and exchange among different LBS. For the second goal, we use semantically rich and expressive models, like UML [4], while for the first one, we adopt the long-praised method of ontologies, realized in the open source, ontology and knowledge-based editor Protégé [38].

To argue about the design choices and show their applicability, we present examples from two characteristic real-world applications, used as case studies, running in the Athens Metropolitan Area: a tourist LBS, in which travelers are carrying mobile devices [8] [37] and a traffic LBS informing the drivers about troublesome situations and alternative routes [24] [5].

Domain Data includes spatial and temporal concepts captured as the object's *position*, *location*, *movement* and *time*. A systematic study reveals that these four spatiotemporal concepts are common and fundamental in all LBS, whether it is, for example, a traffic or a tourist LBS. Thus, it is crucial to share and exchange their semantics. This is achieved by analyzing and model the characteristics and relations of these spatiotemporal concepts. For this, we propose the use of the well-known and long-praised method of ontologies [18] [14] [39], focusing on the comprehension, registration and design of Domain Data. In order to easily realize the ontology, we use the Protégé tool [38]. However, the use of the Protégé tool is just a prototypical one. Any other tool, or standard language such as DAML+OIL [7] would do for this representation.

Special care of location is taken. Until now, all LBS are based on the crude assumption that the location of the mobile object (e.g., a car or a tourist in our case studies) can be simply unambiguously determined; that is, it is *always known* and in *absolute measures*. However this is neither true nor sufficient. In some cases, the position of a moving object is not known, such as when the GPS device is shadowed. In other cases, the user not only cares about her absolute position, but also her surroundings. For example, when tourists visit an archaeological site, the location that matters to them is the actual position in terms of coordinates, as well as a circular 'shape' *around* their current position, which 'includes' items of interest. To capture these semantics, we propose a clear distinction between *location* and *position*. This serves also the need for a better representation, exchange and integration of location from multiple sources, an open problem and a challenge in LBS [26].

Content Data depends on the specific application we are dealing with, e.g., for a tourist LBS this data includes historic facts, restaurant and hotel information. In this work, we model an excerpt of the Content Data existing in a tourist and a traffic LBS by using ontologies for the former LBS and the UML technique for the latter LBS. In the case of the traffic LBS, this leads to the definition and organization of a Moving Object Database (MOD) which includes trajectories, vehicles, routes and their relations and serves as the backbone of the Athens traffic management system. Dealing with these two different LBS scenarios and by using different techniques (i.e., ontologies and UML) shows the diversity of Content Data and consequently the different semantics and design needs. Moreover, it argues that our modeling choices are not tied to specific technology, models and tools.

The third data category comprises *Application Data*, capturing the user profile and service data for the two characteristic application examples. Ontologies are used to denote the data and their semantics. In an LBS scenario, relevant services are discovered, by matching the respective service description with user profiles.

Modeling the semantics of the three data categories leads to the creation of three ontologies: the Domain Ontology, the Content Ontology and the Application Ontology. This structure serves as the backbone architecture to support LBS based on ontologies, with special focus on autonomy and share. To summarize, the contribution of the paper is threefold:

- The categorization of data involved in LBS, based on (data) semantics and use.
- The clear distinction between *location* and *position* in LBS, which solves ambiguities and makes assumptions clear.
- The creation of ontologies (i.e., Domain, Content and Application) for LBS, to represent, share and exchange the concepts of location, position, movement and time among location-based applications.

The rest of the paper is organized as follows: Section 2 gives related, characteristic work focusing on capturing the semantics of LBS. Section 3 presents the types of LBS applications and the Domain, Content and Application Data. Section 4 focuses on the Domain Data and the fundamental related concepts. A clear distinction between *position* and *location* is given; the temporal dimension is treated in a similar way. Section 4 further argues for the use of ontologies in the semantics representation and presents examples in Protégé. Section 5 deals with the Content Data of LBS; it discusses the traffic content data and models their semantics in UML, focusing on the organization of MOD for the traffic management system, while the content data of the tourist LBS are captured in Protégé with ontologies. Section 6 models the Application Data for both the traffic and the tourist LBS with ontologies, and Section 7 concludes this research effort.

2 Related Work

To the best of our knowledge, literature on capturing LBS semantics is quite limited; work has mostly been concentrated on issues related on how to deliver information to the mobile user, rather than what information and semantics are delivered. However, the presence of *location* and *time* play a central role in LBS, and this calls for more rich and complex semantic modeling techniques to capture data involved in the requested services.

In the few existing proposals ([46] [47] [48] [49] [33] [35] and [44]), the use of ontologies has been adopted for this purpose, and quite understandable so, since literature shows many efforts, in other research areas (e.g., biology or business), in which ontologies are used for the analysis and representation of semantics of information.

Ontologies can capture the semantics of information, can be represented in a formal language, and can also be stored to related metadata, thus enabling a semantic approach to information integration. There are several ontology languages as, e.g., compared in [15] and tools to represent ontologies [28] [6] [7]. Protégé [38] is the most popular ontology-editing environment and has been used in many applications, such as medical systems, gene ontology, and business systems.

Furthermore, already, a wide range of applications, such as geographic and biological, call for techniques flexible enough to capture their particularities with respect to space; yet formal ones. [11] proposes a framework for the development of geographic applications by using ontologies. In [11] the reader can also find a systematic review on literature on the use of ontologies in GIS. [3] identifies the role of ontologies in capturing spatial uncertainty. [12] and [13] present methods to bridge the gap between conceptual schemas and ontologies in Geographic Information Systems. Finally, [13] presents the Ontology-Driven Geographical Information Systems framework (ODGIS), which uses ontologies for the comprehensive usage of ontologies for classification purposes, focusing on integrating different kinds of geographic information.

There are some arguments about how useful ontologies are. [20] advises against using ontologies as just a fancy name denoting the result of activities like conceptual analysis and domain modeling [12]. Our position is that ontologies are built to model the semantics of a domain and represent, share and exchange knowledge, while data models and conceptual modeling focus on organizing explicit data and contents resulting in a database. Section 4 elaborates further on this.

Work on ontologies and LBS includes [46] [47] [48] [49] [33] [35] and [44]. They are all based on the assumption that the location is a point with known coordinates.

[46] describes issues involved in supporting an ontology-based information searching process in LBS. It presents an example scenario and gives an architecture based on ontologies that is to be adopted to support share and autonomy in LBS. [49] proposes a collaborative framework for location-based information management consisting of the Query Engine, the Profile Manager, the Data Handler, the TOP Hits Repository, the Data Repository and the Adding Filter. This framework makes it possible to obtain information from heterogeneous sources and improve the request-response efficiency. [33] proposes an ad-hoc model to locate correlative data stores and exchange similar information within a specific community. The model is composed by Data Handlers, Data Stores and proxies and uses ontologies to deal with the spatial relationships between the moving objects. Its continuation [35] proposes the use of ontologies for the management of services in LBS. The proposal exhibits similarities to the newsgroup approach in that both 'systems' are examples of semantic search engines based on user interaction. [44] gives a modular ontology architecture to support different existing ontologies and metadata standards for the web services in Olympia 2008.

The user profile plays an important role in LBS. [47] proposes a profile-based approach to improve the efficiency of the LBS, based on a relational database. As a next step, [48] proposes a way to accommodate user profile needs by using domain and content-depended ontologies. It also suggests the multi-layered abstraction method to organize and present data related to profiles. In this framework, [22] describes a system, which delivers various types of information to mobile devices based on the location, time and profile of the end user. The Event Notification technique has been adopted to trigger actions.

[23] proposes a semantic location model for navigation in mobile environments. It is a hierarchical model and captures connectivity and hierarchical relationships. Again, the assumption here is that location is a point with known coordinates. [42] deals with different types of locations, the way to compute them and to present them. Although the work presented there gives a first taxonomy and general directions about how to handle location in LBS, there is no typical way to categorize them, model them and communicate them with a formal technique, such as a model, ontologies or mathematical representation.

Finally, at the level of services, it is important to point out the effort in achieving an open location services platform (http://www.openls.org/).

3   Types of LBS Applications and Categories of Data

The domain of LBS applications is large and diverse. Here, we present the types of applications supporting location-based services, and analyze the categories of data involved.

3.1   LBS Applications

The GSM Alliance Service Working Group [19] has defined the following types of traditional LBS:

– Emergency Services	– Routing to Nearest Enterprise
– Emergency Alert Services	– Roadside Assistance
– Home-zone billing	– Navigation
– Fleet Management	– City Sightseeing
– Asset Management	– Localized Advertising
– Person Tracking	– Mobile Yellow Pages
– Pet Tracking	– Network Planning
– Traffic Congestion Reporting	– Dynamic Network Control

As LBS, we consider any application involving moving objects and providing services based on positional, temporal and, many times, user profile[1] information. This definition supports the GSM categorization. The position of the object is usually pro-

---

[1] Some works in literature (e.g., [22]) consider the presence of space, time, and user profile *mandatory* to define an LBS; other information, such as 'history' may also exist. However this restrictive definition contradicts the LBS categories of the GSM Alliance Service Working Group; for example the Traffic Congestion Reporting service does not require user profile information. Without affecting the validity and applicability of our results, we chose not to consider mandatory the user profile information.

vided by a mobile device, such as a GPS, carried on/by the moving object or, in the rougher cases, by using positioning in cellular networks.

Two real-world, characteristic, and very different LBS applications are used as case studies in this work:

- a tourist information system providing services based on tourist's location, time and profile [8]. The tourist is equipped with a handheld device having GPS capabilities. Consider, for example, the scenario in which he/she is in the archaeological site of Acropolis and asks 'give me the history of Parthenon', or 'what is the closest monument to me?' or 'what artifacts were found here?'. A tourist LBS should provide answers to these queries. Furthermore, the tourist should be able to provide a profile or preferences, and get information relevant to his interests. For example, a user visiting Acropolis might be interested only in information related to Acropolis and the Pericleus era, i.e., [495BC-429BC].

- an LBS system for traffic management, in which vehicles are equipped with GPS devices [24]. The driver can ask questions such as: 'based on my position, where is a traffic jam?', or 'if there is a traffic jam in the next 10 km, give me alternative routes' or 'give me suggestions to go from Athens to Piraeus'. In this example, it is clear that the user profile is not mandatory, since the user might not have any preferences.

For simplicity purposes and without affecting the validity of our results, we assume that, in the two aforementioned applications, we deal with moving-point objects, i.e., the absolute position of the person or the vehicle that moves is a point.

3.2 Categories of Data in LBS

An important task when building a system, is the analysis and comprehension of the categories of data involved in it, i.e., the related concepts, semantics and use. This helps not only in providing, later on, the appropriate techniques to model and communicate these data, but also to accurately understand the requested services and meet the system's requirements.

Analyzing and comprehending the system data is a modular process: first, we realize the dominating types of data and then the more specific ones. Here, we present three data categories, and show their interconnections; Sections 4, 5 and 6 analyze each one of these categories, give examples and elaborate on the interconnections among them.

In LBS applications, we distinguish three categories of data:

Domain Data. It includes the concepts that are present and characterize all LBS applications.

The common factor behind all LBS is the spatial and temporal dimension, and thus, Domain Data includes fundamental spatial, temporal and spatiotemporal concepts. A careful look across different LBS described by the GSM Group, shows that objects position, location, movement, as well as *time* characterize all of them, whether, for example, we talk about a tourist LBS or a traffic LBS.

Position and location are two terms full of assumptions, which are often used interchangeably. However, in LBS, there is the need to express special meanings and semantics with respect to space. In some cases position refers to absolute coordinates, for example, a car on a road network, while in others, what matters is a greater surrounding area, for example, a tourist wants to know the closest restaurants within 200km radius around him, or the spread of a traffic jam. To better capture semantics, we chose to distinguish these two cases: in reality, the spatial dimension introduces *two new concepts in LBS*: objects' *position* and *location*. Similar issues hold for the temporal dimension; thus, we chose to use the concepts *timestamp* and *time horizon*. A systematic analysis of these concepts reveals different semantics and relations with the environment (Section 4).

Content Data. It is the actual data of the specific application we are dealing with. For example, for the tourist LBS, Content Data is the monuments, the parks, restaurants, etc., while for the traffic LBS is the traffic data, route, and others.

Content data can be: (a) descriptive, for example, restaurant names, or description of museums, (b) spatially-referenced, indicating *where* the actual information is located, seen, or recorded, for example, the location of a museum, and (c) temporally-referenced, showing *when* the information is located, seen or recorded in the system, for example, the time of the traffic jam.

Application Data. It includes the subcategories:

- Profile Data, characterizing the user and the device he is carrying. This can be:
    - (i)   user profile, capturing the user and its preferences. For example, a user visiting Acropolis may be a tourist or a scientist, indicating different interests.
    - (ii)   device profile, characterizing the mobile device the user is carrying, for example, CPU capability, memory characteristics, screen size.
- Service Data, which corresponds to tasks to be accomplished in the specific LBS application. For example, provide specific services to tourists, or to drivers.

Figure 1 illustrates the data categories and subcategories. The lower left corner gives the concepts (in the case of Domain Data) and examples (in the case of the Application and Content Data). The Application and Content Data depend on the specific application we are dealing with. For example, if the application is tourist services over the mobile phone, then the user profile is the one of the tourist with services like 'information about surrounding restaurants' and the content ontology has organized data about restaurants, museums etc.

The three data categories are interconnected, since, for example, in order to provide the service 'closest restaurant' a reference to the restaurants index is needed (i.e., Content Data) and to the position of the user (i.e., Domain Data). There is no association between Content and Domain Data. The Content refers to specific geographic information (for example, location of restaurants in Athens), but this is general spatial data, outside the Domain, which refers to the *where* and *when* the user is.

4   The Domain Data

Domain data characterize all different LBS described by the GSM Group. In fact, the spatial dimension appears in terms of the object's *position* and *location*, the temporal dimension appears in terms of the *time* a the desired information is located, seen or requested, and both dimensions participate in object's *movement*. Next, a systematic analysis of these concepts is presented.

4.1   Semantics of Domain Data in LBS

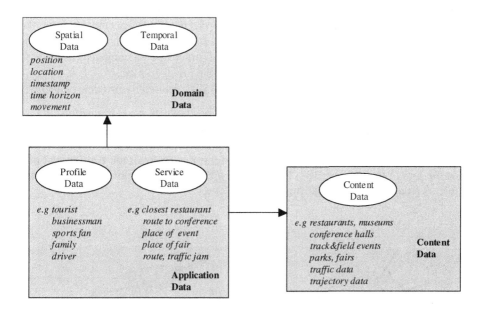

Fig. 1. Categories of data in LBS

A. The spatial dimension introduces the concept of *position* is full of ambiguities and assumptions. Almost in all LBS, there is the crude assumption that the position of a moving object can be simply unambiguously determined [26]; that is, it is *always known* and in *absolute measures*. However this is neither true nor enough. In some cases, for example, the position of a moving object is not known, such as when the GPS device being on it is shadowed.

In other cases, the concept of position has different meanings and values depending on the *application domain* of LBS, and thus cannot be determined by a single notion, it is not unambiguous; thus, it cannot be captured and represented by a unique method or technique. Consider the example of the traffic LBS, in which vehicles are equipped with GPS devices. The absolute, current, *position* of the car is the (x,y) coordinates transmitted by the GPS. However, what really matters to the system to predict and bypass a traffic jam is not just the *position* itself but also the 'shape' or area of 10km *ahead* in the road network, given the fact that the jam is present there.

Analogously, in the tourist LBS providing services to tourists visiting an archaeological site, the *location* that matters to them is the actual (x,y), as well as, the circular 'shape' *around* their current coordinates, which 'includes' items of their interest.

The aforementioned examples are just some of the many we experience everyday indicating that when location *matters*, it is not, only and always, in absolute numbers, but it further depends on the domain of the application.

Moreover, recently, the need to aggregate positions from multiple sources becomes more and more emerging [26]. This is based on the facts that (a) a person may be associated with numerous tracking devices simultaneously, e.g., GPS device on a phone, in a car, etc, and (b) the tracking devices are not always accurate, or may be shadowed as said before, and thus do not deliver the right signal. In this case, the notion of *position* has more than one value and in order to be aggregated, it needs to be analyzed, captured and represented in all possible involved forms.

It is our thesis, that the spatial dimension in LBS is captured by *two new concepts*:

- The absolute (x,y) coordinates of the moving object, which we call **position**
- the 'surrounding area' of position, which we call **reach**. The shape of reach can vary: in the first example (i.e., area of interest in the tourist LBS) it is a circle with a given, predefined, radius. In the second one, it is a shape of an oval (spread of a traffic jam in the traffic LBS). The position and reach of the moving object constitute its **location**. Figure 2 illustrates the two cases.

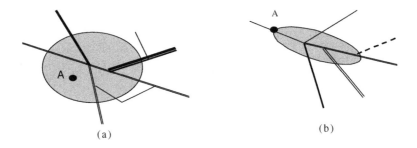

Fig. 2. Different shapes of reach in space: (a) a circular reach surrounding the position A close to a road network, (b) an oval reach ahead of position A on a road network

B. In LBS, it is not only the *where* but also the *when* that matters. Thus, location is related to time. For example, in a tourist LBS it is important to know *when* the tourist is in a location since many facilities depend on that (e.g., when shops are open, etc). Similarly, in traffic LBS for the prediction of a traffic jam, the time a car approaches specific areas matters as rush hours are usually troublesome and matter.

Time, analogously to location, is captured by:

- a *timestamp* t indicating when an action or event happens.
- a *time horizon*, indicating the time period the event or the action has still an effect. For example in the case of a moving vehicle in the traffic LBS trying to avoid a jam, what matters is not only the current time but also the time horizon *ahead* in which the jam will evolve.

Moreover, due to imprecise information, inaccurate measures or device errors, *position*, *reach*, *timestamp* and *time horizon* can have uncertainty, which is usually expressed by the deviation from the accurate value.

C. A fundamental concept in LBS is the *movement* of the object. Movement is defined in terms of position of the moving point object and time, and depending on the application needs, it includes some basic concepts, such as:

- heading, which shows the heading of the moving object
- distance, which gives the distance from the previous position
- direction, which shows the angle to the previous position
- duration, which shows the duration traveled from the previous position.

Figure 3 illustrates this design decision.

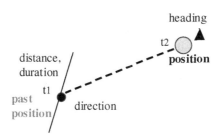

Fig. 3. A diagrammatic description of movement

D. Additionally, the location, i.e., reach and position of the moving object, might be related to others with spatial relationships, which are either topological (e.g, meet, intersect etc) [10], directional (e.g., left, right, etc) or metric, which show distance.

Furthermore, there are applications in which users refer frequently to specific locations for which, neither care nor know their absolute coordinates. For example, the notion of 'work' and 'home' are obvious to everyone and a reasonable service is to be able to 'send me all my SMS as approaching *work*' or to 'download mail going home'. This calls for the concept of *virtual position* related to the user and not to the coordinates. A known user-centered model is comMotion [25]. The approach we propose here captures also virtual positions as it relates position (and location) to the surrounding environment with spatial relationships (e.g., 'approaching' is captured with direction and distance).

4.2 Using Ontologies in LBS

Since Domain Data captures the spatial and temporal dimension and is present in all LBS, the concepts of position, location, time (i.e., time horizon and timestamp) and movement are fundamental and common is all location based applications. This calls for interoperability among LBS, and there is an emerging need to:

– share a common understanding of these spatiotemporal concepts
– make LBS assumptions that exist in literature about these concepts explicit
– exchange and enable reuse of them.

To achieve these goals, we propose the use of ontologies. Here, we focus on the comprehension, registration and design of the aforementioned spatial and temporal concepts. First, we discuss the concept of ontologies and present existing tools and languages supporting them. Then, we show the use of ontologies in LBS, by, initially, using them to represent Domain Data.

About Ontologies. An ontology is an explicit specification of conceptualisation [17]. Ontologies have been long-praised for their efficient use in the comprehension, representation, exchange, share, and integration of domains and concepts [18] [14] [39] [41]. They have been widely used in the past years to describe in an abstract, but accurate way, concepts shared and exchanged among different users, systems, or even people using oral communications. While in the philosophical fields an ontology is the science of being, in the Artificial Intelligence area it is used to describe an engineering, formally defined artifact with specific vocabulary using a set of assumptions regarding the intended meaning of the vocabulary words. Using ontologies to build applications can help avoid problems, such as inconsistency and poor understanding among communicating parties.

The Artificial Intelligence literature contains many definitions of ontology. Many of them contradict each other. Generally speaking, in the engineering world, an ontology is a formal and declarative representation which includes the names for referring to the terms in that subject area and the logical statements that describe what the terms are, how they are related to each other, and how they can or cannot be related to each other. Ontologies therefore provide a vocabulary for representing and communicating knowledge about some topic and a set of relationships that hold among the terms in that vocabulary.

In practical terms, the design of an ontology includes:

– the definition of classes or concepts in the ontology
– the arrangement of the classes in a taxonomic (subclass-superclass) hierarchy, if it exists
– the definition of properties and the description of the allowed values for these properties
– the definition of restrictions on the values of the properties, such as cardinality

An ontology, together with a set of individual instances of classes with specific values of properties, constitutes the knowledge base of the application.

The line between ontologies and conceptual schemas is thin. One could argue that the process of creating ontologies is conceptual modeling. Another approach is that using conceptual models to represent a domain of the application is adequate; proposals like that do exist [16] [41]. However, besides the fact that ontologies offer more flexibility in information representation, there are differences between conceptual schemas and ontologies:

- at the schematic level, an ontology is usually a forest of diagrams, while a conceptual schema –based on the strict literature definitions– is not, and
- ontologies are used to exchange and share common information (for example, the 'location', 'position' and 'time') among applications belonging to the same domain (for example, fleet management, mobile services etc., are all LBS), while conceptual schemas are used to model data in one application.

Some proposals about ontology definition include also the definition of rules to add semantics and to infer knowledge. Rules represent implicit knowledge about classes and their relationships. If one adopts this ontological approach, then this is one more difference between ontologies and conceptual schemas as rules exist only in ontologies.

One way is to see ontologies as an abstraction of conceptual schemas. Overall, ontologies are semantically richer than the conceptual schemas as they are built for different purposes: the former to represent a domain in a knowledge base, and the latter to represent contents of a database.

There are several ontology languages and tools, which are used to build ontologies. The most popular of them are compared in [15]. DAML+OIL [7] is the standard ontology language and close to the standards developed by W3C [45]. Chimaera [6], Ontolingua [28] OntoBuilder [27] and Protégé [38] are some of the most known tools as ontology editing environments.

In literature, there are also proposals for the structure of ontological environments. A representative one is [20] which structures an ontology to sub-ontologies: (a) the upper ontology, which includes abstract and philosophical issues, (b) the domain ontology which includes specific domains, such as tourism, weather, (c) the task ontology which contains knowledge about the usage, and (d) the application ontology which combines and extends the knowledge of all other ontologies. Depending on the application domain, several ontologies can be identified at the levels listed above.

Ontologies in LBS. As the need for capturing more semantics in LBS is growing together with the demand of structured information and services, domain experts started using ontologies in location-based applications. The design of ontologies is a modular task, i.e., it is important to define their structure and their interconnections, starting from the global or more dominant ones and then the more specialized ones, creating in this way, a structure, or an architecture. Moreover, more and more libraries of ontologies *do* exist today, such as the DAML ontology library (available at www.daml.org) or the Ontobuilder [27] (available at http://ie.technion.ac.il/OntoBuilder). This gives the expert the ability to acquire ontologies from different environments; however, it is crucial for integrity reasons to categorize them at the right level of the ontology architecture.

Some works follow specific architectural proposals for LBS. [44] follows the architecture presented in [20] to present an ontology list for semantic GeoServices for

Olympia 2008. [46] presents a different architecture to share ontologies in LBS but also keep their autonomy. The elements in this architecture are: (a) the global ontology, (b) the local ontologies, which correspond to local sources, (c) the shared ontologies, (d) the mediator and (e) the integrated ontology.

In our work, we propose an LBS architecture whose structural components follow this rationale and support the data categories of Figure 1 of Section 3. Thus, we propose the design of the:

- Domain Ontology, thus consisting of the Space Ontology and Time Ontology,
- Content Ontology, and
- Application Ontology, consisting of the Profile Ontology and Service Ontology.

This ontology architecture (a) keeps the Domain Ontology independent of other semantics and characteristics, and thus it can easily be exchanged and shared among LBS, according to Section 3, and (b) respects the, well-documented in literature, separation between applications (i.e., services and profiles) and context (i.e., information). The following sections describe the aforementioned ontologies.

4.3   The Domain Ontology of LBS

After the systematic analysis, clarification and relation of the fundamental concepts of the Domain Data of LBS, we proceed on the design of the Domain Ontology. We use Protégé[2] for the design and, further, the full development of the Domain Ontology of both the tourist and the traffic LBS.

Protégé [38] has (a) a graphical and easy-to-use interface, (b) a flexible knowledge model, and (c) an extensible plug-in architecture. With respect to Section 4.2, it includes:

- classes, which are the modeled concepts
- slots, which represent first-class objects representing properties or attributes of classes. A slot can be of an atomic type (e.g., float, integer, etc) or if an instance type, which means that it is an instance of another class.
- facets, which are constraints on allowed slot values, such as cardinality, defaults, allowed classes and others.
- axioms, which specify additional constraints

The distinction between classes and instances is not an absolute one. Both individuals and classes themselves can be instances of classes [38]. The main advantages of Protégé are that:

- It is easy and understandable enough for the domain expert to use it to develop the ontologies of his interest.
- It is an adaptable tool, which we can tune to support new languages and formalisms quickly. This is important as on the one hand, a number of new semantic-web languages and representation formalisms are emerging, but on the other, there is no agreement made yet.

---

[2] Protégé, as of Feb. 15, 2003, is available in version 2.0.

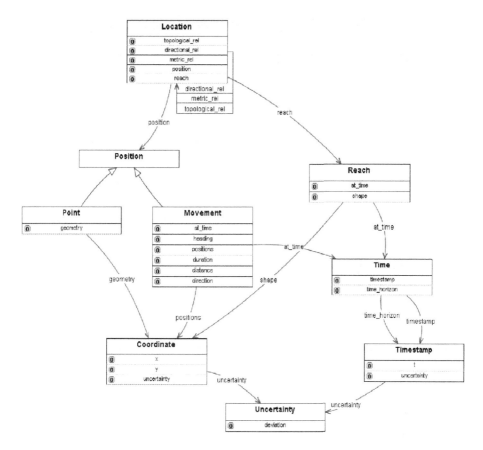

Fig. 4. Classes and relations among them in Protégé, capturing the Domain Ontology of LBS

- It can be used for the development and management of ontologies and applications today without waiting for standards.
- The supported model is an open and extensible one, allowing for plug-ins serving specific purposes.

The output of the design on Protégé can be expressed in widely used semantic web languages, such as RDF (Resource Description Framework), XML, Ontology Inference Layer (OIL), and JDBC which support the share and exchange of the designed data, in our case, the Domain Data.

However, we should make clear, that the use of Protégé is a prototypical one; any other tool or language with equal expressive power would do for this purpose. For this, we do not attempt to present specific implementation details that depend on the particularities of the ontology-editing tool, but rather use it as an illustration for the concepts we discuss.

Figure 4 illustrates the classes, instances, and slots among them, capturing the Domain Ontology of LBS in Protégé. In a class, when a slot is not of atomic type (e.g.,

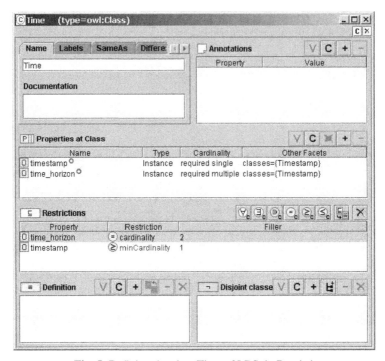

Fig. 5. Defining the class Time of LBS, in Protégé

float, integer, etc), but it is an instance of another class then it is depicted with an arrow. In order to explain Figure 4, for reasons of semantic simplicity and comprehension, whenever a slot of a class A is of type instance of class B, we say it that the class A is 'related to' class B.

Location is related to *Position* and *Reach*, while it has *topological_rel*, *metric_rel* and *directional_rel* relations with other *Location*s.

Movement and *Point* are subclasses of *Position*. *Movement* has as slots *heading*, *duration*, *distance* and *direction*, which are of atomic type and *positions* and *at_time* which are instances of classes *Time* and *Coordinate*, and thus depicted as relations.

Point has as slot *geometry*, which is an instance of class *Coordinate* with slots *x* and *y*, capturing the coordinates of the moving point object and *uncertainty*, which is an instance of *Uncertainty* class capturing the *deviation* from the true value.

Reach has as slots *at_time* and *shape*, which are instances of *Time* and *Coordinate*, respectively.

Time has as slots *time-horizon* and *timestamp*. *Timestamp* has as slots *t*, which is of atomic type, and *uncertainty*, which is an instance of *Uncertainty*.

Figure 4 translates to forms, such as the one presented in Figure 5. Figure 5 illustrates the definition of the class Time. The timestamp and time_horizon are instances of Timestamp and their cardinality is defined. time_horizon has cardinality of 2 as it needs two timestamps to be defined. As seen, Figure 5 allows for the definition of details, specifications and restrictions, such as disjoint classes, documentation and others.

The Domain Ontology (cf. Figure 4) of LBS includes:

- the Space Ontology, including the classes of *Location*, *Position*, *Point*, *Reach*, *Coordinate*, *Uncertainty* and *Movement*
- the Time Ontology, including the classes of *Time*, *Timestamp*, *Uncertainty* and *Movement*.

Protégé has a plug-in to import ontologies from other ontology editing environments, as for example, Ontolingua [30] or DAML [7]. The Time Ontology for example is available in the Ontolingua ontology library [29], as Simple-Time, including time-points and time-ranges and following the Allen's time theory [1]. Thus, one would argue, about how useful is to develop new ontology and not importing existing ones from available servers; this approach has been used in other application domains [36]. The Time Ontology we define here includes the movement class and thus is tailored to the needs of LBS. For reasons of integrity the Allen's relation should be included to relate the Time instances (as it happens with the Space instances). However, since this is trivial, we do not present it.

Similarly, the Space Ontology is also available in existing libraries. Again, just its adoption and import to the LBS knowledge base is not enough as the specific slots and instances we design and use are explicit for LBS and absolutely necessary to capture the very specific semantics of LBS.

5   The Content Data and Ontology of LBS

Following Figure 1, the next step is the analysis of the semantics of Content Data and its representation. We present excerpts of the Content Data from both the traffic and the tourist LBS. We chose two different approaches in capturing Content Data for the two LBS:

- for the traffic LBS we chose to use the UML technique in order to organize the huge amount of traffic data. This results to the Moving Object Database (MOD), which includes trajectories, vehicles and routes in UML, which was chosen due it is popularity, high-degree of comprehension and expressiveness. MOD is the core of a traffic management system on which, in many application environments [24], data mining functions are applied to extract information about traffic prediction. The reader can find more details about MOD in [5].
- for the tourist LBS we stayed focused on the use of ontologies, resulting to the Content Ontology.

By using different techniques to capture the semantics of the Content Data, we show the diversity of LBS Content Data and consequently the different semantics and design needs. Furthermore, this builds on the fact that our design choices are not bounded to the use of specific technology and tool.

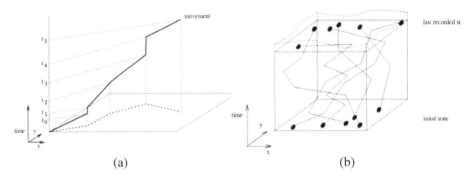

(a)   (b)

Fig. 6. Moving point objects: (a) a trajectory and (b) several trajectories in evolving in a finite region

5.1   Content Data of a Traffic LBS

The organization of traffic Content Data in the Moving Object Database (MOD), calls for a further, more detailed and in-depth understanding of objects, their properties and relations[3] related to the concept of movement.

Movement in a Traffic LBS. Consider the following scenario using a traffic management system to monitor the traffic flow in its city area of Athens, Greece. By monitoring the movement of specific vehicles (e.g., delivery trucks, public transport, taxis, etc.) one can ask the following queries: 'find the vehicles that just entered Athens', or 'find the vehicles that left Athens an hour ago,' or more general 'find locations with a larger number of vehicles' (i.e., typical traffic jam pre-condition). Representing such moving objects as point objects their movement can be illustrated as shown in Figure 6. The solid line in Figure 6(a) represents the movement of a point object. Space (x- and y-axes) and time (t-axis) are combined to form a 3D-area. The dashed line shows the projection of the movement in two-dimensional space (x and y coordinates).

In order to record the movement of a vehicle, we need its position at all times, i.e., on a continuous basis. However, GPS and telecommunications technologies only allow us to sample an object's position, i.e., to obtain the position at discrete instances of time such as every few seconds. By, later on, interpolating these samples, we can extract the movement of the object. The simplest approach is to use linear interpolation, as opposed to other methods such as polynomial splines [2]. The sampled positions then become the end points of line segments of polylines, and the movement of an object is represented by an entire polyline in three-dimensional space. In geometrical terms, the movement of an object is termed a *trajectory*; in other words, trajectory is the trace of the vehicle in time.

Figure 6(b) shows a spatiotemporal space (the cube in solid lines) and several trajectories (the solid lines) contained in it. Time moves in the upward direction, and the top of the cube is the time of the most recent position sample. The wavy-dotted lines on top symbolize the growth of the cube with time.

---

[3] In the classical database meaning.

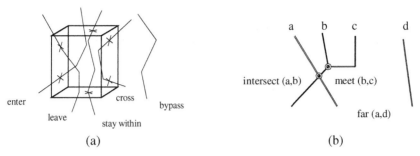

Fig. 7. Relationships: (a) trajectory/spatial environment and (b) trajectory/trajectory

The trajectory representation is adequate to derive certain properties and relations of the object's movement:

Properties in MOD. Trajectories are characterized by a set of different properties depending on the application requirements. Some of the most common properties are:

- the speed of the movement (indicated by the inclination of the trajectory)
- the heading of the vehicle,
- the covered area, indicating the area the vehicle covered during its trip,
- the traveled distance, and
- the traveled time.

Based on our studies [32] [34], the aforementioned representation is adequate for mobile database modeling, since it gives answers to simple questions, such as 'which area did vehicle A-4592 cover during its trip?' and to more complex ones, like 'which vehicles left Athens after midnight moving East and were found close to each other 2 hours later?'.

Relations in MOD. Through their movement, trajectories relate to their environment in different ways over time. In the following, we discuss to types of relationships, namely how a trajectory can relate to its (spatial) environment and to other trajectories.

Relations between a trajectory and its spatial environment. Trajectories can have relations with the spatial environment. which includes other spatial objects. These can be either infrastructure elements, such as roads, parks, buildings, etc. but also imaginary entities such as city boundaries or query regions. In the temporal context these spatial entities become three-dimensional (i.e., space and time dimensions) represented by e.g., a 3D region. We distinguish five basic relationships (Figure 7(a)), but others can also be included:

- *stay within*, when the trajectory is all the time in the range of interest,
- *bypass*, when the trajectory passes by the range of interest,
- *leave*, when the trajectory leaves the range of interest,
- *enter*, when the trajectory enters the range of interest,
- *cross*, when the trajectory crosses the range of interest.

Relations among trajectories: Additionally, relevant positions among trajectories need to be registered at time points. The most common ones based on topological reasoning [10] are the following (Figure 7(b) depicts four of them):

- *intersect*, indicating that two trajectories intersect,
- *meet*, showing that two trajectories touch at one point
- *equal*, when two trajectories coincide,
- *near*, when two trajectories are close to each other, based on definitions on what 'close' means
- *far*, when two trajectories are away from each other.

Note that the concepts of far/near are context sensitive and thus depend on the application domain. For example, what is 'near' for two airplanes is rather 'far' for two cars and even farther for two pedestrians.

Having defined the above, one can ask for trajectory(-ies) fulfilling one or more conditions; from the simple 'which area did vehicle X cover during its trip?' to the more complex 'which vehicles left this area after midnight moving East and were found close to each other 2 hours later?'

Some of the aforementioned properties and relations have been also presented in Section 4.1 to capture the semantics of movement and location. Here, for the needs of MOD, we present them in more detail. Additionally, for both types of relations there exists a substantial amount of work in literature with respect to the way of how two real world objects are topologically associated. In this work, we just include the fundamental ones.

Organizing MOD for a Traffic LBS. The various concepts relating to trajectories presented in the previous section are organized to define the underlying data model of a MOD. Following the well-known methodology of a database design, including the phases of conceptual modeling, logical modeling and implementation, we initially use conceptual modeling to capture the semantics of the aforementioned concepts in an organized manner. For the conceptual representation, we use the class diagram of UML [4] due its popularity, high-degree of comprehension and expressiveness.

Figure 8 illustrates the conceptual schema of MOD and exhibits five major classes, namely, trajectory, 3D-region, vehicle, road, and road segment and two relations which are modeled as object classes: the relation among trajectories ('trajectory/trajectory') and the one between trajectory and 3D-region ('trajectory/environment'). Due to the fact that movement, changes continuously other properties of the objects involved in the database, such as the speed of the vehicle, the heading of the vehicle, and relations among them, such as far (i.e., the two vehicles are far'), or near (i.e., the two vehicles are near') it is essential to capture functions or operations on objects. For example, 'GetSpeed' shows the speed of the vehicle at a given time point, or 'Far' gives a boolean answer about whether or not two vehicles are far from each other. An operation is a service applied on an object. The UML class diagram proved to be expressive enough to capture all the aforementioned elements and semantics.

To capture a 'trajectory', we need an identification of the mobile device (indicated by 'object id'), the actual trajectory ('trajectory id') as well as the position of the

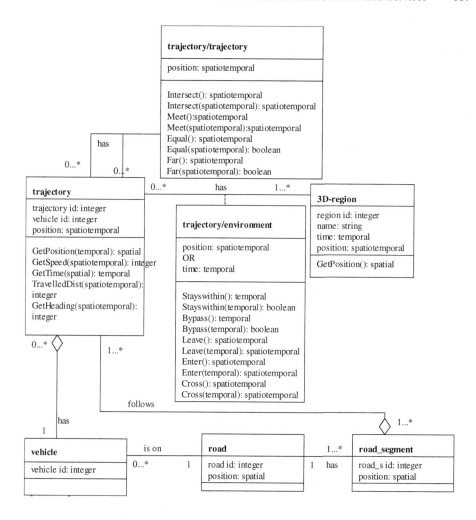

Fig. 8. An excerpt of the database schema of MOD

trajectory itself. In other words, 'position' describes the trace of the moving vehicle. The data types used are abstract, since they only should indicate the dimensionality of the parameter. More concrete instances of data types can be found in, e.g., [21]. A set of operations, e.g., GetSpeed(spatiotemporal), GetTime(spatial), and TravelledDistance(spatiotemporal), GetHeading(spatiotemporal) are prototypical and show what type of information can be derived from the trajectory data, e.g., to compute the traveled distance or the heading of a trajectory, we apply an operation that uses a spatiotemporal range as a parameter.

The '3D-region' class is prototypical to denote the spatial environment of the trajectory (as part of a 3D-region representing the 2D-space and the time dimension (cf. Figure 1(b))). It is a fundamental object class of MOD. As stated previously, the 3D-region can be built up as time progresses and the objects move; in this case it shows the total covered area.

Trajectories 'have' (one or more) relations either with other trajectories, or their 3D-region class. Figure 8 contains the respective classes functions to compute such relationships. E.g., 'Leave' without parameter computes the spatiotemporal positions at which a trajectory left a given instance of a 3D-region class. To restrict the operation, we can use an argument to the function. In the case of Leave it is a temporal argument, i.e., the search for spatiotemporal positions at which the trajectory has left the region is restricted to a given time interval. In the class 'trajectory/environment' the parameter 'position' or 'time' capture the result of the function. Equally, so does 'position' in relation 'trajectory/trajectory'.

Finally, for reasons of integrity, we capture the obvious object classes 'vehicle', 'road' and 'road segment'. Note that the relation between vehicle and trajectory is an aggregation, as one vehicle can appear and disappear (due to loss of the GPS signal) and thus its route is a combination of trajectories. The same happens between 'road' and 'road_segment'.

Figure 8 depicts, as a prototypical example, only the basic classes; other classes, for example 3D-lines (e.g., road-networks in time) that exhibit different relations with moving objects (e.g., moving along, etc.) can also be accommodated in this approach.

The rationale and choices presented here have the main advantage of describing two basic concepts: (a) *the trajectory* of the moving object by keeping track of its movement, and (b) the moving object itself, by recording its last known position. The spatiotemporal framework in which the movement takes place can either be built on the fly (i.e., while objects move) or be pre-defined (e.g., Athens in a specific time interval).

5.2   The Content Ontology of a Tourist LBS

Tourist content data includes information relating to entertainment, museums, history, etc. These Content Data can be structured in the form of an ontology/taxonomy grouping the content into a hierarchical set of categories of data.

Figure 9 shows an excerpt of a Content Ontology in Protégé showing related classes in a tourist LBS. The arrow indicates a superclass-class relation; e.g., History is the superclass of Historic_Event and Historic_Site.

Such a taxonomy can be seen as a general means to structure content data. For example, for the tourist LBS, given the fact of the historic battle of Marathon, which happened in the year 490BC, this information can be categorized under the 'Historic_Event' class. Figure 9 illustrates this by having the instance 'Battle_of_Marathon' connected to the respective class by a dashed arrow. Another example is the historic site 'Acropolis' categorized under to the 'Historic_Site' class.

The organization of Figure 9 exhibits similarities with other existing taxonomies existing, such as the dmoz.org open directory [9].

Content is related to spatial information in terms of the position of the facilities it describes. For example, the content of a tourist LBS includes the positions of all restaurants in Athens. This spatial data is not related to the Space Ontology (Section 4.3), as the later describes the whereabouts of the moving object.

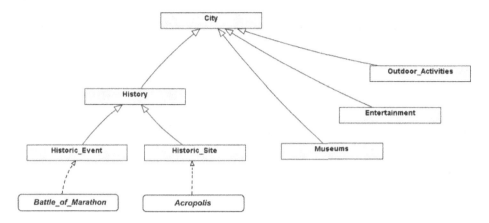

Fig. 9. An excerpt of the Content Ontology for a tourist LBS

6   The Application Ontology of LBS

Following Figure 1, the next step is the analysis and the modeling of the Application Data. Application Data (cf. Figure 1) refers to profiles and services, related to specific applications, in our case, the traffic and tourist LBS. In this section we discuss the types of user profile data and services and possible ways to represent them in the Application Ontology (cf. Section 4.2). We do not deal with the device profile (cf. Section 3.2), as its data is governed by the characteristics of the particular device and this is outside the scope of this work.

6.1   The Profile Ontology

Users do have preferences with respect to what information they usually request, and considering mobility, as to when and to where they do this. Recording these data leads to creating a *user profile*. It represents the choices and the needs of each individual user so that (a) the mobile device behaves in a way desired by the user and (b) information of interest is *forwarded* to the user in both *synchronous* (pull) and *asynchronous* (push) modes. In both cases the *position* of the user and the *time* are essential features and are taken into account. For example, in the tourist LBS, the user profile changes depending on the position of the user (e.g., 'when I am in Berlin, my profile is *business*, when in Bahamas, my profile is *tourism*) or even on the time (e.g., 'after 8pm receive only information about entertainment').

The user profile can be: (a) explicitly defined by the user and (b) implicitly be modified by a data mining module that takes the demographic data of the user and his/her behavior patterns into account, where behavior patterns can be categorized into (i) spatiotemporal behavior (i.e., the user motion patterns in space through time) and (ii) previous choices that the user has made regarding information access.

Figure 10 gives an example of a simple, explicitly defined User Profile Ontology, for the traffic and the tourist LBS, that structures the interested of a 'User' based on

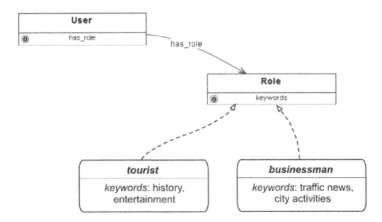

Fig. 10. An excerpt of the user profile of the Application Ontology for the tourist LBS

the concept of a 'Role.' Roles are aggregations of profile specifications with the interests specified as keywords for each role. E.g., if a user "activates" his 'tourist' role, he wants to be notified of services relating to history and entertainment (cf. Figure 10). His 'businessman' role states his interest in traffic news and city activities. As we will see later on in Section 6.3, this explicit specification of interests can be used for automatic service discovery.

6.2 The Service Ontology

Since services rely on data, relating to the content ontology of Figure 9, a similar ontology can be derived to structure services in relation to the data they provide. Services have the spatial dimension, in the sense that they are structured analogously to the *Site*[4] they refer to. For example, for the tourist LBS, the Entertainment, Museums, History, and Outdoor Activities, which are all services, are related to specific sites. The same holds for a traffic LBS, in which the Traffic, i.e., Traffic_Jam and Traffic_Load always refer to specific sites.

Figure 11 illustrates an excerpt of the Service Ontology of the traffic and the tourist LBS.

Site can be a point or an area providing specific services, e.g., facilities in the area.. A service ontology is used to discover services based on a request. A request is specified in terms of the spatial parameter, i.e., *Site*, and additional descriptive information such as keywords. Using the service ontology, all services will be structured according to their respective spatial scope, i.e., their *Site*, and the semantic category they belong to. Matching a request to an actual service is done by matching the *location* (of the user) and the keywords characterizing the request to the respective *Site* a service covers and the specific category a service belongs to, respectively. Matching keywords onto categories can be done by, e.g., measuring the word distance between the set of keywords and the matching category descriptions in the taxonomy [43][40].

---

[4] We use the term *Site* to avoid confusion with *location* and *position*, which are reserved words in this work.

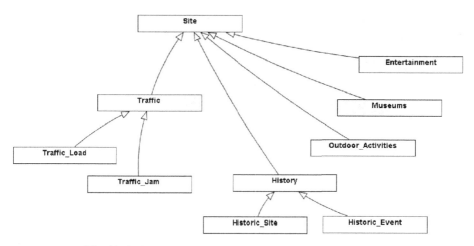

Fig. 11. A simple Service Ontology for a traffic and tourist LBS

Consider the scenario from the tourist LBS in which a service providing historical information relating to Acropolis, by using the classes of Figure 11, is categorized under the 'Historic_Site' class. The typical example 'give me the history of the place I am' (cf. Section 3) relates, spatially, the *location* (in this case *position*) of the tourist with the area he is in.

Analogously, the example from the traffic LBS, 'based on my position, where is a traffic jam in 5km ahead', relates, spatially, the location (cf. Figure 2b) to the area of a traffic jam.

The spatial relation between *Site* and *location* is achieved by using well-known spatial relationships [10].

6.3 Automatic Service Matching

In this section we discuss some ways services are provided by using concepts from the Domain Ontology, and the Application Ontology.

Besides searching for services based on explicit requests (pull), services can be triggered implicitly (push) by matching a user profile onto service descriptions by using agents [31]. Assuming, in a tourist LBS, an extended user profile (cf. Section 6.1) contains information about preferences a person has when she is traveling as a tourist, e.g., history and entertainment (cf. Figure 10). By traveling to a new destination at some point his 'tourist' profile will be matched onto available service description and, e.g., a service related to museums information will be presented to him. On the other hand, when he is on the job, which is that of a traveling salesman by car, his profile specifies that he is interested in traffic-related services. In this case, reaching a new destination, traffic-related services will be presented to him.

Events (other than profile declaration) trigger this service discovery. *Position* and *location* play a central role. It acts as a trigger to send related information to the user. For example, when a tourist interested in history reaches the Athens city centre, a service is activated that presents information about the Athens archaeological museum.

Another factor that can act as a trigger for service discovery and activation is the *history* of the user, which can be registered in his profile. For example, in the tourist LBS, considering a tourist who frequently visits museums. Then, even if he has not defined explicitly 'history' or 'museums' as a preference in his profile, when coming to Athens he still will be presented services that inform him about the Athens archaeological museum. A detailed discussion about the role and use of events in LBS can be found in [22].

7 Conclusions

In this work, we analyze, comprehend and model data semantics of LBS. The analysis of data leads to the *Domain, Content* and *Application Data* categories depending on the related concepts and their use. To model these data categories we adopt the semantically rich UML as well as the long-praised method of ontologies, depending on the application needs and the complexity of semantics.

Modeling the semantics of the three data categories leads to the creation of three different ontologies: the Domain Ontology, the Content Ontology and the Application Ontology. This structure, illustrated in Figure 12, serves as the backbone architecture to support LBS based on ontologies, with special focus on autonomy and share.

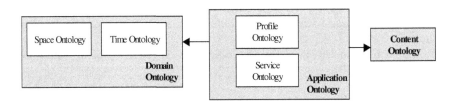

Fig. 12. The Ontology structure of LBS

The three data categories are interconnected; the Application Ontology is connected to both the Domain Ontology and the Content Ontology, since the Service Ontology relates to the Content and to the Space and the Time Ontology, and the Profile Ontology relates to the Space and the Time Ontology. For example, in order to provide the service 'closest restaurant' a reference to the restaurants index is needed and to the position of the user. There is no association between Content and Domain Ontology. The Content refers to specific geographic information (for example, location of restaurants in Athens), but this is general spatial data, outside the Domain Ontology, which refers to the *where* and *when* the user is.

A major contribution of this work, is the creation of ontologies (i.e., Domain, Content and Application) for LBS, to represent, share and exchange the concepts of location, position, movement and time among location-based applications.

Another important strength of this work is the clear distinction between *location* and *position*, which solves ambiguities and makes assumptions about these two concepts clear. Distinguish *location* from *position* further helps on the accurate semantic modeling and the representation, exchange and integration of location from multiple sources.

Finally, the applicability and feasibility of our design choices are shown with examples from two real case studies, the tourist and the traffic LBS for the Athens Metropolitan Area.

Acknowledgements

This work is supported by the *DB-Globe: A Data-centric Approach to Global Computing* project, funded by the European Commission under contract number IST-2001-32645, and the *IXNHΛΑΤΗΣ: A Traffic Management System* project, funded by the Hellenic General Secretariat of Research and Technology. The authors wish to thank EMFASIS Telematics, Greece, for providing real data for the case studies.

References

1. Allen, J.: Towards a General Theory of Action and Time. Artificial Intelligence 23 (1984) 123-154.
2. Bartels, R.H., Beatty, J.C., and Barsky, B.A.: An Introduction to Splines for Use in Computer Graphics & Geometric Modeling. Morgan Kaufmann Publishers, Inc. (1987).
3. Bittner, T., and Winter, S.: An Ontology in Image Analysis. In Proc. of Integrated Spatial Databases: Digital Images and GIS (1999) 168-191.
4. Booch, G., Rumbaugh, J., and Cobson, I.: The Unified Modeling Language User Guide. Addison-Wesley (1999).
5. Brakatsoulas, S., Pfoser, D., and Tryfona, N.: Modeling, Storing and Mining Moving Objects Databases. In Proc. of the International Database Engineering and Applications Symposium (IDEAS) (2004) 68-77.
6. Chimaera Ontology Environment, Web page, www.ksl.stanford.edu/software.chimaera (2004).
7. DAML.Org. Reference Description of the DAML+OIL ontology markup language. Web page http://www.daml.org/2001/03/reference (2004).
8. DBGlobe: A Data-centric Approach to Global Computing. IST Project, Project number IST-2001-32645.
9. DMOZ.org: Open Directory Project. Web page http://www.dmoz.org (2004).
10. Egenhofer, M.: Reasoning about Binary Topological Relations. In Proc. of the 2nd Symposium on Spatial Databases (SSD) (1991) 143-160.
11. Fonseca, F.: Ontology-driven Geographic Information Systems. In Proc. of the 7th ACM Symposium on Advances in Geographic Information Systems (1999) 14-19.
12. Fonseca, F., Davis, C., and Camara, G.: Bridging Ontologies and Conceptual Schemas in Geographic Information Systems. GeoInformatica 7(4) (2003) 355-378.
13. Fonseca F.T., Egenhofer, M.J., and Agouris, P.: Using Ontologies for Integrated Geographic Information Systems. Transactions in GIS 6(3) (2002) 231-257.

14. Frank, A.: Spatial Ontology: A Geographical Point of View. Spatial and Temporal Reasoning, Kluwer, Dordrecht (1997) 135-153.
15. Garshol, L.M.: Topic Maps, RDF, DAML, OIL – A comparison. Web page http://www.ontopia.net/topicmaps/materials/tmrdfoildaml.html (2001).
16. Granefield, S., and Purvis, M.: UML as an Ontology Modeling Language. In Proc. of the Workshop on Intelligent Information Integration, 16th International Joint Conference on Artificial Intelligence (IJCAI) (1999) 46-53.
17. Gruber, Th.: A Translation Approach to Portable Ontology Specifications. Knowledge Systems Laboratory – Stanford University, Stanford, CA, Technical Report KSL 92-71 (1992).
18. Gruber, Th.: Toward Principles for the Design of Ontologies Used for Knowledge Sharing. Int. Journal of Human-Computer Studies, Vol. 43 (1993) 907-928.
19. GSM Alliance Services Working Group, Web page http://www.gsmworld.com/about/structure/serg.shtml (2004).
20. Guarino, N.: Formal Ontology and Information Systems. In Proc. of FOIS'98, Trento, Italy. Amsterdam IOS Press (1998) 3-15.
21. Guetting, R., Böhlen, M., Erwig, M., Jensen, C. S., Lorentzos, N., Schneider, M., and Vazirgiannis, M.: A Foundation for Representing and Querying Moving Objects. ACM Transactions on Database Systems 25(1) (2000) 1-42.
22. Hinze A., and Voisard A.: Location- and Time-based Information Delivery in Tourism. In Proc. 8th International Symposium in Spatial and Temporal Databases (SSTD) (2003) 489-507.
23. Hu, H., and Lee, D.: Semantic Location Modeling for Location Navigation in Mobile Environment. In. Proc. Of the IEEE International Conference on Mobile Data Management (MDM) (2004) 52-61.
24. ΙΧΝΗΛΑΤΗΣ: A Traffic Management System. Research Academic Computer Technology Institute, 2002-2004. ENTER Project, Hellenic General Secretariat of Research and Development (2004).
25. Marmasse, N., and Schmandt, C.: A Location-Aware Information delivery with comMotion. In Proc. 2nd International Symposium on Handheld and Ubiquitous Computing (HUC) (2000) 157-171.
26. Myllykaki J. and Edlund S.: Location Aggregation from Multiple Sources. In Proc. of the 3rd International Conference on Mobile Data Management (MDM) (2002) 131-138.
27. OntoBuilder Project. Web page http://ie.technion.ac.il/OntoBuilder (2004).
28. Ontolingua: Ontolingua System Reference Manual. Web page http://www-ksl-svc.stanford.edu:5915/doc/frame-editor/index.html (2004).
29. Ontolingua: Ontolingua ontology library, Web page http://www.ksl.stanford.edu/software/ontolingua/ (2004).
30. Ontolingua: Ontolingua Server. Web page http://www-ksl-svc.stanford.edu:5915/ (2004).
31. Panayiotou, C., and Samaras, G.: Personalized Portals for the Wireless User: An Agent Approach. Journal of ACM/Baltzer Mobile Networking and Applications (MONET), special issue on "Mobile Commerce" (2004).
32. Pfoser, D., Jensen, C.J., and Theodoridis, Y.: Novel Approaches in Query processing for Moving Objects. In Proc. of the conference on Very Large Data Bases (VLDB) (2000) 395-406.
33. Pfoser D., Pitoura, E., and Tryfona N.: Metadata Modeling in a Global Computing Environment. In Proc. of the 10th ACM International Symposium on Advances in Geographic Information Systems (ACMGIS) (2002) 68-73.

34. Pfoser, D., and Theodoridis, Y.: Generating Semantics-Based Trajectories of Moving Objects. International Workshop on Emerging Technologies for Geo-Based Applications, Ascona, Switzerland (2000) 59-76.
35. Pfoser, D., Tryfona, N., and Verykios, V.: Services-based Data Management in a Global Computing Environment. In Proc. of the 3rd International Workshop on Web and Wireless Geographic Information Systems (W2GIS) (2003) 45-53.
36. Pinto, S., and Peralta, D.: Combining Ontology Engineering Subprocesses to Build a Time Ontology. In Proc. of the International Conference on Knowledge Capture (2003) 88-95.
37. Pitoura, E., Abiteboul, A., Pfoser, D., Samaras, G., and Vazirgiannis, M.: DBGlobe: a service-oriented P2P system for global computing. SIGMOD Record 32(3) (2003) 77-82.
38. Protégé Project. Web page http://protégé.stanford.edu (2004).
39. Smith, B., and Mark, D.: Ontology and Geographic Kinds. In Proc. of the International Symposium on Spatial Data Handling (SDH) (1998) 308-320.
40. Sycara, K. P., Widoff, S., Klusch, M., Jianguo Lu: LARKS: Dynamic Matchmaking Among Heterogeneous Software Agents in Cyberspace. Autonomous Agents and Multi-Agent Systems 5(2) (2002) 173-203.
41. Tryfona N. and Pfoser D.: Designing Ontologies for Moving Objects Applications. In Proc. of the International Workshop on Complex Reasoning on Geographic Data, Paphos, Cyprus (2001).
42. Ubicomp Workshop: Location Modeling for Ubiquitous Computing. Workshop Proceedings, 2001.
43. Valavanis, E., Ververidis, C., Vazirgianis, M., Polyzos, G. and Norvag, K.: MobiShare: Sharing Context-Dependent Data & Services from Mobile Sources. In Proc. of the IEEE/WIC International Conference on Web Intelligence, (2003) 263-271.
44. Weissenberg N., and Gartmann R.: Ontology Architecture for Semantic GeoServices for Olympia 2008. In Proc. Münsteraner GI-Tage, Münster. IfGIprints 18 (2003) 267-283.
45. W3C: OWL Web Ontology Language Reference. Web page http://www.w3.org/TR/owl-ref/ (2004).
46. Yu S., Aufaure M., Cullot N., and Spaccapietra S.: Location-based Spatial Modeling Using Ontology. In Proc. of the 6th AGILE Conference (2003).
47. Yu S., Spaccapietra, S., Cullot, N., and Aufaure M.: User Profiles in Location-based Services: Make Humans More Nomadic and Personalised. In Proc. of the International Conference on Databases and Applications (2004).
48. Yu S., Aufaure M., Cullot N., and Spaccapietra S.: A Collaborative Framework for Location-based Services. In Proc. of the International Conference on Databases and Applications (2004).

Geospatial Conceptualisation: A Cross-Cultural Analysis on Portuguese and American Geographical Categorisations

Paulo Pires

Superior Institute of Statistics and Information Management,
Universidade Nova de Lisboa, Portugal
ppires@isegi.unl.pt

Abstract. In 2001 David Mark and Barry Smith[1] published a study that aimed to establish how non-expert subjects conceptualise geospatial phenomena in the United States of America. This paper contributes to comparing the results from the study performed by David Mark and Barry Smith with a similar study applied to Portuguese non-expert subjects.

In response to a series of questions, differently phrased, 160 non-expert subjects (university students from several parts of Portugal and various academic areas) were asked to give examples of geographical categories such as "Natural earth formation" or "Something that can be portrayed on a map". The answers were used to statistically count the most mentioned terms. The Portuguese results were compared with those of the United States.

Although Portuguese results differ slightly in the number of items presented, the most mentioned terms match with the United States results.

We are therefore able to derive that the conceptualisation of geographical entities of Portuguese subjects is similar to that of United States subjects, e.g. Portuguese subjects also refer to mostly physical characteristics such as trees and mountains.

At a time when interoperability and ontological studies gain importance in the Geographical Information Systems/Science, this ongoing initiative points out the need of integrating trans-border geographical conceptualisations.

Keywords: Cognition, Geographical Information Systems, Cross-Cultural analysis.

1 Introduction: A Cross-Cultural Analysis on Portuguese and American Geographical Categorisations

Since Aristotle, Ontology has been conceived as a branch of metaphysics that studies the theory of objects and their links. Today, ontologies are applied in Information Systems. How can we measure different geographical conceptualisations? Are there cross-cultural differences for the same conceptualisations?

---

[1] www.tandf.co.uk/journals
IEEE Transactions on Knowledge and Data Engineering, 15 (2): 442-456, 2003.

S. Spaccapietra and E. Zimányi (Eds.): Journal on Data Semantics III, LNCS 3534, pp. 196–212, 2005.

"Specification of a conceptualisation" [1] is a very common answer to the question: what is ontology? In philosophy, ontology refers to the subject of existence. When applied to the field of Information Systems it refers to the description of concepts and relations of entities to a certain reality/domain.

But in what way can ontologies be useful to Geographical Information Systems? Ontologies can be a very efficient means to improve the creation of geographical information in order to support human activities [2].

The work presented in this paper is part of the research conducted by the author for his master thesis concerning Portuguese Water Bodies and Ontologies, in which a practical study in the form of a survey was performed. This work was based on similar approaches taken in other parts of the world.

David Mark and Barry Smith, from Buffalo University, were two of the pioneers of the kind of study [3] that is developed in this paper. They took the model survey by Battig and Montague [4], and built another that would allow the construction of an ontology of geographical categories. The subjects of David Mark and Barry Smith's survey were non-experts in the geographical information world.

Using a similar structure as in Smith and Mark [3] and adding some categories to the survey, we built two domain ontologies of water-related geographical concepts [5].

Formal Conceptual Analysis (FCA) [6] allows the fusing and analysis of both ontologies in the form of a lattice diagram. With this we measured the agreements and disagreements of experts' and non-experts' conceptualisations of water body entities. We also measured the differences by the number of nodes in the lattice diagram in similarity to Kokla and Kavouras [7].

The survey also allowed the comparison of some of the results stated in Smith and Mark [3] with the Portuguese results. Parallel experiments were carried out in Finland, United Kingdom and Croatia - see [3] for details. They all concluded similar trends, but the most significant result was when the subjects were asked about geographical features, geographical objects and geographical adjectives: they almost exclusively elicited elements of the physical environment of geographical scale or size, like rivers or mountains.

In this paper we focus our attention on the analysis of the results obtained in the cross-cultural comparison between Portuguese and American results.

2 Ontologies: The Link of Terminological Concepts

Nowadays common sense tells us that Information is relevant to our daily life. Therefore, "it is essential to invest in the creation of languages and tools that allow the sharing and transmitting of Geographical Information through a formal and accessible scheme" [8].

For instance, consider the mathematical definition of a limited function:

$$\lim_{x \to c} f(x) = L \Leftrightarrow \forall \varepsilon > 0, \exists \delta > 0 \; / \; |x-c| < \delta \Rightarrow |f(x) - L| < \varepsilon$$

Every mathematician in the world understands the reading of this formula in exactly the same way. But in a science such as Geographical Information Systems the agreement on the meanings of words and symbols is not so obvious.

Ontologies appear as a means to help solve what may be called the "Tower of Babel" problem. It consists of the fact that when different data (from different sources) is put together, problems of terminological and conceptual incompatibilities arise.

In this context an ontology is a description of entities and their relations to a determined universe/domain. In other words, as Gruber [1] defends, ontology can be seen as "a specification of a conceptualisation".

The conceptualisation mentioned in this definition refers to formalisation, that is, ontology from the philosophical perspective (formal ontology).

This is similar to the study of a mathematical function in which the following function f is studied:

$$f: R\text{->}R$$
$$x \text{->} x$$

Genesereth and Nilsson [9] defined conceptualisation with the same structure as above:

$$<D;R>$$

where D stands for domain (in the above example D=R, that is, the domain of this function belongs to the set of real numbers) and R stands for Relation (in the above example f(x)=x).

Ontologies and Geographical Information Systems may seem two very distant fields, but currently they are used together in order to achieve a unique and complementary goal: a better understanding of what exists around us. In this way ontologies can improve the creation of geographical information in order to support human activities.

3   The Cross Cultural Analysis Between Portugal and USA: Methodological Issues

There are several possible approaches that can be used in order to understand what perception people have of a specific matter. Surveys are the most common method applied by sociologists, psychologists and general researchers in social sciences.

Smith and Mark [3] carried out an experiment on students in order to understand how non-expert subjects conceptualise geospatial phenomena. We applied a similar experiment, in which we gave the subjects six category-titles and asked them to write down five items for each category.

The objective was to determine if Portuguese students and US students (from various universities) have similar conceptualisation of certain geographical entities. To do this, our experiment was conducted in approximately the same conditions.

We applied the survey to 574 students from different colleges but only 161 of these surveys were used for the cross-cultural analysis (the remaining surveys will be used in the follow-up of this study). Figure 1 illustrates the geographical location of the cities where the surveys took place (Size of circles represent the number of non-expert subjects per geographical location).

In an attempt to be representative, the selection of the 161 surveys from the 574 was made by taking 8 from each course (exception in courses with fewer surveys than this), i.e. we had 8 surveys from the mathematics course, 8 from economics.

As in the American case we also applied the survey within a time limit, that is, students had 30 minutes to fill in the entire survey and this was controlled by teachers or the author of the thesis, obliging them to write the first thoughts that came into their heads for each category in approximately 30 seconds *per category* (so that they could give examples for each category).

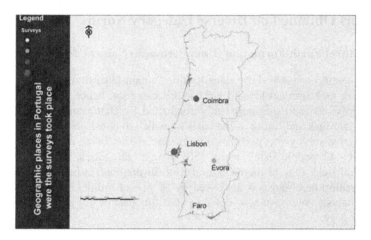

Fig. 1. Geographical locations in Portugal where the surveys took place

In response to a series of different category titles, section I of the survey asks subjects to give examples of geographical categories. This allows us to statistically determine the most mentioned terms and compare them with the American results.

Note that due to some methodological differences in the application of the survey, the results of the two countries can only be compared in ordinal terms and not in cardinal terms (i.e. the order of several examples mentioned, but not the percentual value obtained itself). In effect, we asked our subjects to suggest a maximum of six examples for each category, while in the American experiment subjects were asked to fill a blank sheet of paper for each of the same categories. This different approach partially explains the higher percentual values generally obtained in the American results, regardless of the category.

Another significant methodological difference was that American subjects were given a different question to answer for each category (i.e. there were 5 kinds of questions, corresponding to the 5 categories, and each survey only had one of the following: a kind of geographic feature; a kind of geographical object; a geographical concept; something geographical; and something that could be portrayed on a map). That allowed a comparison of the answers given and, on that basis, to conclude if there were significant differences due to the way in which the questions were asked. On the other hand, in the Portuguese case, the same 5 questions were presented to each subject[2] (all in the same questionnaire), and he or she had to give an answer (also with several items) to each one of them. The objective of this different methodology was to have more robust results in terms of the homogeneity of answers, reducing interpersonal differentiations. However, this can have some secondary effects, namely, because the subjects know the 5 different questions that were posed, they tend to differentiate their own answers to each question more, induced by the variations and tone of the question...Thus, there is expected to be a greater diversity and variability of answers in the Portuguese than in the American case.

4   Results Obtained on Diverse Category Norms

4.1   A Natural Earth Formation (Uma Formação Natural da Terra)

The first question concerned the category title "a natural earth formation".

The Mark and Smith study did not use this category, hence a comparison is made with the previous (and original) study conducted by Battig and Montague; this had been used by Mark and Smith as the basis on which to construct their survey.

The original survey was performed in the late sixties, using a similar methodology; they asked 442 people using many (56) diverse categories, some of them in the geographical field. One of them (natural earth formations) is interesting for the study that we develop here and was also used by Mark and Smith [3] as the starting point for their analysis. We have also opted to start this part of our survey with a question on this category.

We can observe the main results and compare them with those of the US (in this specific case, from the study of Battig and Montague, [4]), in Figure 2 - Results for the category A Natural earth formation.

Subjects named more than 100 different earth formations, mostly related to natural earth creations. On observing the table we see that the mountain was the most mentioned item, with a frequency of 71 individuals (44%), followed by volcano and river.

In Battig and Montague's results, the mountain was also the most mentioned item within this category (401 subjects out of 442) followed by hill and valley. The latter is the 4th most rated item in the Portuguese result.

We can highlight two important conclusions in the comparison of these results. First that there is a great similarity in the results of the two surveys, with 4 common answers in the top 5 (mountain, river, valley, rock), and another two (ocean, lake) in the top 10 answers (and other similar results in the subsequent positions); secondly, there is a greater relative concentration of answers in the top categories in the American case (with higher relative frequencies for the first results e.g. 91% for mountain) than in the Portuguese case (with a greater dispersion of top answers, the first one – mountain – achieving only 44% of the subjects).

The main differences to be noted are that the hill was in 2nd place in US results as opposed to the volcano in the Portuguese results (neither of them in the top 10 of the other); in addition greater importance was given to canyon, cliff and cave in the

---

[2] The study was conducted in Portuguese.

American Results (Valid N=442)	%	Portuguese Results (Valid N=161)	%
Mountain	91	Mountain (Montanha)	44
Hill	51	Volcano (Vulcão)	35
Valley	51	River (Rio)	30
River	33	Valley (Vale)	24
Rock	24	Rock (Rocha)	14
Lake	22	Water (Água)	11
Canyon	18	Ocean (Oceano)	13
Cliff	17	Plain (Planicie)	11
Ocean	17	Sea (Mar)	11
Cave	16	Lake (Lago)	10

Fig. 2. Results for the category "A Natural Earth formation"
Source: own elaboration, based on Battig and Montague's survey (referred in Mark and Smith, 2001) for the US case, and our survey data for the Portuguese case.

American case (possibly as they are more visible in their everyday reality), and to water, sea and plain in Portugal (probably for the same reasons). These results (5 of the top 10 Portuguese results are related to water) as well as the fact that the river appears here as the 3rd most mentioned entity, can be partially explained by the Portuguese reality, i.e. Portugal is a country in which water plays a constant role in every day life and history and it is therefore natural that subjects should mention it. On the other hand, this explanation does not justify the high position given to the volcano, which is perhaps due to the strong media attention attributed to the phenomenon.

Another point to highlight is the fact that Portuguese students mentioned nothing specific, that is,, they mentioned "serra", but not "Serra da Estrela". In the American results subjects mentioned one specific formation, the Grand Canyon (to be precise, 14 subjects). This can imply that both groups mentioned more categories than specifics.

4.2 Something That Could Be Portrayed on a Map[3] *(Algo que possa ser representado num mapa)*

All the following questions assumed the work pattern used by Mark and Smith, in spite of the few methodological differences mentioned above.

---

[3] In order to highlight the differences between Portuguese language/meaning and the American language a table is presented at the end of this chapter with the translated Portuguese answers, Table I.

In this first category ("something that could be portrayed on a map), the River is ranked in 1st position with nearly 40%, followed by Road, City and Country (see Figure 3 – Something that could be portrayed on a map). Again these are very similar to the US results; the first items chosen by the American subjects were River, City, Road, Mountain, Country, Lake, Ocean, State and Continent, all of which correspond to the top 7 answers of the Portuguese with the exception of state and lake (it would be very unlikely for Portuguese subjects to write these given that the State is not an administrative unit in this country, and lakes are not very common.). Although the relative values are (again) much lower than in the US case, the 3 top answers (river, road, city) were the same.

Furthermore, both American and Portuguese subjects mentioned essentially more physical than human geographical characteristics (although in both cases this was the category that presented a relative stronger weight of human/constructed geographical characteristics).

We expected subjects studying Geography, Urban Planning and Architecture to give answers like houses, buildings or Portuguese regional units (e.g. Nut I, Nut II and Nut III), but their most mentioned entities were also geographical characteristics such as river and road (with some exception for the Geography course where they mentioned demographic rates such as population and birth rate).

American Results	Frequency	%	Portuguese results	Frequency	%
Total	51			161	
River	31	60,8	River	64	39,8
City	30	58,8	Road	54	33,5
Road	27	52,9	City	47	29,2
Mountain	25	49,0	Country	34	21,1
Country	23	45,1	Ocean	26	16,1
Lake	21	41,2	Continent	22	13,7
Ocean	18	35,3	Mountain	21	13,0
State	15	29,4	Sea	15	9,3
Continent	12	23,5	Releif	14	8,7
Street	8	15,7	Locality	11	6,8
Town	8	15,7	Topography	9	5,6
Highway	7	13,7	Soil	9	5,6
Park	6	11,8	Altitude	9	5,6
Sea	5	9,8	Plain	8	5,0
Building	5	9,8	Contour Line	8	5,0
County	5	9,8	Declivious	8	5,0
Elevation	5	9,8	Stream	8	5,0

Fig. 3. Results for the category "Something that can be portrayed on a map"
Source: own elaboration, based on Mark and Smith [3] for the US case and on our survey data for the Portuguese case.

To conclude, we note that in this category on the whole both groups mentioned more objects with artificial or man made boundaries (particularly in US), which leads us to conclude that the subjects use maps to show aspects of human geography more than physical geography. Or, as Mark and Smith put it, considering the weight that human contents currently have in the work of geographers, this (relative) weight of human activities and items in these question contrary to all the others, captures the meaning of "geographical" as this term is employed in scientific contexts. It seems that geographers "are not studying geographical things as such things are conceptualised by naïve subjects. Rather, they are studying the domain of what can be portrayed on maps" [3].

4.3 A Kind of Geographical Object *(Um Objecto Geográfico)*

In this category there is not much convergence in the US and Portuguese results in terms of the items indicated by the subjects; however, the two countries both tend to indicate mostly (or at least, a relative high percentage of) examples of small, portable items. Map is the item most commonly mentioned in Portugal (41% in Portugal,) and is highly ranked (3rd with 13%) in US, immediately after the two main expressions referred in nearly all categories in this survey (due to the methodology adopted): River and Mountain). Several other top expressions mentioned (e.g. globe or compass in US; Compass, GPS, astrolabe in Portugal) refer to small items. We believe that this is due to the word *Object* which probably suggests portable things.

The trend is much more marked in the Portuguese case due in part perhaps to the language question, but also and mainly to the conditions in which the survey was applied; subject are more likely to extremate and differentiate the items referred if they have the 5 different questions to answer than if they have just one and do not know all the others.

This, together with the similarities between the two groups can be better understood by looking at Figure 4 - A kind of a geographical object.

American Results	Frequency	%	Portuguese results	Frequency	%
Total	54			161	
Mountain	23	42,6	Map	66	41,0
River	18	33,3	Compass	60	37,3
Map	17	31,5	GPS	18	11,2
Ocean	16	29,6	Astrolabe	15	9,3
Lake	13	24,1	Satellite	13	8,1
Globe	11	20,4	Geodesic Mark	11	6,8
Peninsula	10	18,5	Quadrant	10	6,2
Continent	10	18,5	River	9	5,6
Hill	9	16,7			
Sea	8	14,8			
Compass	8	14,8			
Valley	7	13,0			
Island	7	13,0			
Rock	6	11,1			
Plane	6	11,1			
Land	6	11,1			
Desert	6	11,1			
Country	6	11,1			
Atlas	6	11,1			
Town	5	9,3			
State	5	9,3			

Fig. 4. Results for the category "A kind of geographical object"
Source: own elaboration, based on Mark and Smith [3] for the US case and on our survey data for the Portuguese case.

Note that a compass is the only other common geographical object mentioned in top positions by both sets of enquiries (37% in Portugal, in 2nd place; 14.8% in USA). Curiously, "astrolabe" (an ancient Portuguese invention, from the Discoveries era) is mentioned in 4th place in this country, certainly for its symbolic status in historical and cultural terms, just after the modern GPS (in 3rd).

From the above analysis (and from the comparative results with the other categories studied) we can assume that the subjects' notion of object relates to physical, material things used in geographical matters. In the Portuguese case, subjects would have been expected to answer more natural geographical things like mountain, river or street (as Americans), but due to the methodology of the survey (and perhaps language differences) this above-mentioned assumption is clearly reinforced in the Portuguese case.

4.4 Something Geographical *(Algo Geográfico)*

It could be said that the results in this category continue with this same trend, clearly reinforcing the idea that the results are very similar to those of the US. Portuguese students mentioned Map, River, and Mountain in the top places (see Figure 5 - Something geographical) as something geographical. The US students mention all the same in their top 5 responses.

Furthermore, 6 of the 7 Portuguese top answers are in the 8 American leading positions. The only exception is City (mentioned by the Portuguese), which with Relief (8th position) differs only from Lake and Hill in the American figures.

There is no doubt that this category leads to a much more generalistic approach to the geographical contents in both countries. As in the US, the Portuguese subjects also mentioned the same items here that they had mentioned in the other categories, confirming the fact that it is the most generalistic category of all. The aspects indicated here tend to be the same as the most popular ones in other categories, essentially related to physical or artificial aspects (river, mountain, city, lake) or representation methods (map), but not specific items to this category. That is particularly evident in the Portuguese case (without any new reference mentioned more in this category than in any other, to the data selected) but also in the US case (only with the generic issues land and the world referred more in this category than in any other).

4.5 A Geographical Concept *(Conceito Geográfico)*

As in the US survey, Portuguese subjects also had difficulty answering this category because it was here that we observed the highest number of missing values. The results are very varied in both countries which reinforces that conclusions.

Simultaneously, this is the only category in which there are no common answers in the selected top responses from both countries. In all the 13 expressions mentioned by at least 10% of the Americans and the 7 expressions referred by at least 5% of the Portuguese (criteria used to establish the tops presented in these tables) there is not a single common expression.

American Results	Frequency	%	Portuguese results	Frequency	%
Total	51			161	
Mountain	32	62,7	Map	20	12,4
River	26	51,0	River	18	11,2
Lake	25	49,0	Mountain	17	10,6
Ocean	18	35,3	City	16	9,9
Hill	11	21,6	Ocean	13	8,1
Map	11	21,6	Country	13	8,1
Sea	9	17,6	Sea	9	5,6
Country	8	15,7	Relief	8	5,0
Continent	8	15,7			
Island	7	13,7			
Plateau	6	11,8			
Desert	6	11,8			
Land	6	11,8			
Plane	5	9,8			
Volcano	5	9,8			
Forest	5	9,8			
Peninsula	5	9,8			
The world	5	9,8			
City	5	9,8			

Fig. 5. Results for the category "Something geographical"
Source: own elaboration, based on Mark and Smith [3] for the US case and on our survey data for the Portuguese case.

American Results	Frequency	%	Portuguese results	Frequency	%
Total	54			161	
Mountain	23	42,6	Latitude	32	19,9
River	19	35,2	Longitude	26	16,1
Ocean	16	29,6	Altitude	19	11,8
Sea	11	20,4	Population	11	6,8
Lake	10	18,5	Population Density	11	6,8
Continent	9	16,7	Natality	10	6,2
Plateau	8	14,8	Localitation	9	5,6
Valley	7	13,0			
Island	7	13,0			
Map	7	13,0			
Peninsula	6	11,1			
Delta	6	11,1			
Land	5	9,3			

Fig. 6. Results for the category "A geographical concept"
Source: own elaboration, based on Mark and Smith [3] for the US case and on our survey data for the Portuguese case.

Portuguese subjects mentioned more (in this category) both demographic and cartographic concepts (Latitude, Longitude, Altitude, Population, Population density and Birth - see Figure 6 - A geographical concept) while US subjects mentioned things like Mountain, River, Ocean, sea or Lake more frequently in the category (note that these main/top choices of the Americans are essentially geographical formations, more than specific concepts).

These differences can also result either from language specificities or from the methodology used in the survey, with a much clearer based pattern towards "conceptual" items in the Portuguese case, where people were confronted with the other possible answers.

As mentioned above, we can also read from the results in both countries that subjects had difficulty deciding what a geographical concept is, assuming this category as "larger" and generic one.

4.6 A Kind of Geographic Feature *(Característica Geográfica)*

Finally, the last category concerned a "geographical feature". Here the results are again very different in relation to the concrete expressions mentioned, although again with similarities in the pattern with which subjects considered the category used.

The difference in concrete expression suggested may be explained not only by the referred trend towards extreme differences in the Portuguese case due to the methodology applied but also (particularly in this case) by the relevant distinctions in language use for this category. In effect, feature is used in a much broader sense for geographical purposes in

English than the Portuguese translation ("característica"), which immediately suggests "characteristics" or "properties" of an object/subject. It is normal that Portuguese answers are more centred on conceptual geographical items than on concrete geographical features.

There is no doubt that Portuguese subjects link geographical features/ characteristics with cartographical concepts (see figure 7-A geographical feature), because the most mentioned items were Latitude, Longitude and Altitude. Some (mostly from architecture) also mentioned demographic concepts, but this was not so frequent. On the other hand, American subjects referred mostly to the physical environment: mountain, river, lake and ocean in this category.

American Results	Frequency	%	Portuguese results	Frequency	%
Total (N)	54			161	
Mountain	48	88,9	Altitude	46	28,6
River	35	64,8	Latitude	45	28,0
Lake	33	61,1	Longitude	44	27,3
Ocean	27	50,0	Relief	28	17,4
Valley	21	38,9	Climate	13	8,1
Hill	20	37,0	Plain	13	8,1
Plane	19	35,2	Declivious	11	6,8
Plateau	17	31,5	Profundity	10	6,2
Desert	14	25,9	Distance	9	5,6
Volcano	10	18,5	Plateau	8	5,0
Sea	9	16,7	Contour line	8	5,0
Island	8	14,8			
Peninsula	8	14,8			
Forest	6	11,1			
Stream	6	11,1			

Fig. 7. Results for the category "A kind of geographic feature"
Source: own elaboration, based on Mark and Smith [3] for the US case and on our survey data for the Portuguese case.

Thus, this difference can probably be explained by the question of translation, assuming that Portuguese subjects have a broader understanding of the term (characteristics), than the one usually adopted by the Americans when referring to what they call the specific "geographical features" of a place (something noticeable in a particular area of a country like a river, a hill or a valley).

Thus, this difference can probably be explained by the question of translation, assuming that Portuguese subjects have a broader understanding of the term (characteristics), than the one usually adopted by the Americans when referring to

Table 1. English and Portuguese terms

Portuguese answers (English/Portuguese)

English	Portuguese
Altitute	Altitude
Astrolabe	Astrolábio
City	Cidade
Climate	Clima
Compass	Compasso
Continent	Continente
Contour Line	Linha de costa
Country	País
Slope	Declive
Geodesic mark	Marco geodésico
Latitude	Latitude
Locality	Localidade
Location	Localização
Longitude	Longitude
Map	Mapa
Mountain	Montanha
Birth rate	Natalidade
Ocean	Oceano
Plain	Plano
Plateau	Planíce
Population	População
Population density	Densidade Populacional
Quadrant	Quadrante
Relief	Relevo
River	Rio
Road	Estrada
Satelite	Satélite
Sea	Mar
Soil	Solo
Stream	Riacho
Topography	Topografia

what they call the specific "geographical features" of a place (something noticeable in a particular area of a country like a river, a hill or a valley).

In this category, our expectation of finding similar items mentioned to those in the "geographical concept" category was confirmed. This might mean that subjects make a narrow distinction between concept and characteristic. Subjects did not associate characteristics like small, long, salty, i.e. generic characteristics, at all. We assume they had interpreted geographical characteristic as geographical terminology.

5   Conclusion: Similar Trends Between Portuguese and US Results?

Many of the conclusions from the above analysis can be drawn from the observation of figures 8 and 9 which show the previous results on a comparative basis in a concise fashion.

A selection of the expressions found in the answers from both countries was used to construct these tables. Due to the large number of expressions that resulted from those answers, some criteria had to be used to select expressions that facilitated the comparison amongst categories. Smith and Mark [3] decided to concentrate their analysis only on terms mentioned with a statistically (more) significant frequency and arbitrarily chose to study only terms that were listed by at least 10% of the subjects for at least one of the five phrasings. Similarly, we opted for the same solution. However, as the dispersion of answers at lower levels was greater in the Portuguese case (by survey characteristics), we decided to admit only terms that were listed by at least 5% of the subjects for at least one of the five categories as criteria for study …

If we compare the items highlighted in the various tables (representing, for each expression the category in which it stands out most), we can have a general notion of which items people (relatively) associate most to each category, and therefore, the notion which subjects have of that category.

This method (central to the Mark and Smith analysis) allows us to subscribe the generality of results outlined above. So what conclusions can be drawn?

First, American and Portuguese subjects (despite a number of differences in the concrete expressions indicated due to territorial, cultural and even language specificities) have some similarities in the way they approach the various categories. In spite of a relevant discrepancy in the way both look at "geographical features", all the remaining categories demonstrate many common points in the understanding both sets of subjects have of each category ("geographical objects", "geographical concepts", "something geographical", "something that could be portrayed on a map", as well as the Battig and Montague test category "natural Earth formation").

Secondly, the differences amongst categories are much more clearly marked in the Portuguese survey, which results (at least, partially) from the methodology used in the application of the survey. While in the American case, the students were confronted, alternatively, with only one of the 5 categories/questions, the Portuguese were asked to answer all categories/questions in each survey. This difference naturally implies that the subject tends to differentiate the answers to the various categories, resulting in greater contrast.

With the results presented in this work we conclude that in effect, Portuguese and US students in general <u>do</u> have similar conceptualisations of the categories presented, differing slightly in the kind or number of items presented (particularly in some specific points mostly related to the physical and historical-cultural characteristics of their own country); the most mentioned items generally match.

This can be partially explained by the fact that both Portuguese and American samples are university students. Therefore we are speaking (as mentioned at the start of this work) of a specific group of the population with common characteristics in both countries. Indeed, although the subject of the degree differs in the studies of the two countries, the main body of the university program is the same as well as the main socio-economic characteristics of this population group.

On the other hand, we can also probably conclude that another explanation for this might be that we are dealing with a universal concept of "geography" (or universally

	Feature	Object	Something	Concept	Map	Total
Total	54	56	51	51	51	263
Mountain	48	23	32	23	25	151
River	35	18	26	19	31	129
Lake	33	13	25	10	21	102
Ocean	27	16	18	16	18	95
Valley	21	7	4	7	0	39
Hill	20	9	11	3	0	43
Plane	19	6	5	4	1	35
Plateau	17	4	6	8	0	35
Desert	14	6	6	4	0	30
Volcano	10	4	5	3	0	22
Island	8	7	7	7	3	32
Forest	6	4	5	1	3	19
Stream	6	2	2	3	1	14
Map	0	17	11	7	0	35
Globe	0	11	4	0	0	15
Peninsula	8	10	5	6	1	30
Compass	0	8	0	1	2	11
Rock	1	6	3	2	0	12
Atlas	0	6	2	2	0	10
Land	2	6	6	5	0	19
The world	0	0	5	1	3	9
Sea	9	8	9	11	5	42
Delta	4	1	0	6	0	11
City	1	4	5	0	30	40
Road	1	2	3	1	27	34
Country	2	6	8	4	23	43
State	0	5	3	1	15	24
Continent	1	10	8	9	12	40
Street	0	1	1	1	8	11
Town	0	5	2	0	8	15
Highway	1	0	0	0	7	8
Park	0	0	0	0	6	6
Building	0	1	0	0	5	6
County	0	2	0	0	5	7
Elevation	0	0	0	1	5	6

Fig. 8. Most frequent terms for the American subjects
Source: own elaboration, based on Mark and Smith [3] for the US case and on our survey data for the Portuguese one.

	Feature	Object	Something	Concept	Map	Total
Total	161	161	161	161	161	161
altitude	46	0	6	19	9	80
latitude	45	2	5	32	6	90
longitude	44	0	4	26	4	78
relief	28	2	8	3	14	55
climate	13	0	5	1	1	20
plain	13	4	4	2	8	31
declivious	11	0	1	4	8	24
profundity	10	0	0	4	1	15
distance	9	0	2	5	1	17
contour line	8	2	2	2	8	22
plateau	8	0	4	3	3	18
map	0	66	20	6	1	93
compass	0	60	1	0	0	61
soil	2	30	1	1	9	43
gps	0	18	4	2	0	24
astrolabe	0	15	0	1	0	16
satellite	0	13	1	0	0	14
geodesic mark	0	11	1	0	0	12
quadrant	0	10	0	0	0	10
population	4	4	4	11	5	28
populational density	3	0	1	11	5	20
natality	1	1	4	10	0	16
localization	6	0	1	9	2	18
river	1	9	18	3	64	95
continent	1	4	5	3	22	35
mountain	7	8	17	2	21	55
ocean	0	3	13	2	26	44
country	1	7	13	1	34	56
city	0	7	16	1	47	71
road	0	4	1	0	54	59
stream	1	2	2	0	8	13
sea	2	1	9	0	15	27
locality	0	1	2	0	11	14
topography	2	0	2	0	9	13

Fig. 9. Most frequent terms for the Portuguese subjects
Source: own elaboration, based on Mark and Smith (2001) for the US case and on our survey data for the Portuguese case.

geographical concepts); in other words, in certain aspects at least they may be scarcely influenced by the socio, cultural and economic composition of the population (wealthy, ethnicity, sex, nationality, and so on).

Note also that even the effective geographical differences (e.g. there are neither deserts nor active volcanoes in Portugal) seem to be a determinant factor for the different answers.

It would be very interesting to expand this analysis by comparing this study with others conducted in different realities. Namely, the studies performed in Finland, Croatia and U.K (see [3]), would be an interesting contribution to this debate.

For now, however, our conclusion must be that we cannot assume that Portuguese and Americans generally have different conceptualisations of these geographical concepts.

Equally, and in more practical terms, we believe that the importance of this work also lies in the ongoing implementation of the European Water Frame Directive in National laws, as this initiative points out the need to integrate trans-border geographical conceptualisations; this specific work can contribute to finding a possible methodology for other studies within European countries.

As Mark and Turk [10] said "It is important to raise questions such as: Are there significant cross-cultural and cross-linguistic differences in the way human beings perceive and cognize their environments at geographical or landscape scales?" Indeed, the results of these questions are of importance to those who conceive and design Geographical Information Systems.

Acknowledgment

We would like to acknowledge the valuable help of Professor Werner Kuhn and Professor Marco Paínho that made this project possible.

References

1. Gruber, T.: What is an Ontology, available on www-ksl.Stanford.edu/kst/what-is-an-ontology.html (2001)
2. Kuhn, W.: Ontologies in Support of Activities in Geographic Space, International Journal of Geographical Information Science, vol. 15, n° 7, (2001) pp. 613-631
3. Smith, B. and D. Mark: Geographical categories: an ontological investigation, International Journal of Geographical Information Science, vol. 15, n° 7, (2001) pp. 591-612
4. Battig, W. F. and W. E. Montague: Category norms for verbal items in 56 categories a replication and extension of the Connecticut Norms, Journal of Experimental Psychology Monograph, 80, Part 2, (1968) pp. 1-46
5. Pires, P. and Brox, C.: Measuring semantic differences between experts\' and non-experts\' conceptualisations. published at the GEOPRO 03 - International workshop semantic processing of spatial data, Mexico City, Mexico (2003)
6. Ganter, B. and R. Wille: Formal Concept Analysis: mathematical foundations, Dresden: Springer (1999)
7. Kokla, M. and M. Kavouras: Fusion of top-level and geographical domain ontologies based on context formation and complementarity, International Journal of Geographical Information Science, vol. 15, n° 7, (2001) pp. 679-687
8. Mota, L., J. Bento and L. Botelho: Ontologias de Informação Geográfica, Comunicação apresentada ao ESIG2001, Oeiras, Taguspark, Novembro (2001)
9. Genesereth, M. R. and N.J. Nilsson: Logical Foundation of Artificial Intelligence, Los Altos, CA: Morgan Kaufmann (1987)
10. Mark, D., Turk, A.: Landscape categories in Yindjibarndi:Ontology, Enviroment, and Language, published at the COSIT2003, Ittingen, Switzerland (2003)
11. Carnap, R.: Empiricism, Semantics and Ontology, Supplement to Meaning and Necessity, A study in Semantics and Modal Logic, enlarged edition, Chicago: University of Chicago Press (1956)
12. Gruber, T.: A translation approach to portable ontologies, Knowledge Acquisition, 5(2), (1993) pp. 199-220
13. Guarino, N.: Formalizing Ontological Commitments, in Bishr, Y. and W. Kuhn (Eds.), The role of ontology in modelling geospatial features, Münster: Institut für Geoinformatik (1999)

14. Rodriguez, A. and M. Egenhofer: Determining semantic among entity classes from different ontologies, IEEE Transactions on Knowledge and Data Engineering, 15 (2): (2003) pp. 442-456
15. Smith, B.: Ontology, in Kim, J. and E. Sosa (Eds.), A Companion to Metaphysics, Cambridge / Oxford: Basil Blackwell, (1995) pp. 373-374
16. Smith, B. and D. Mark: Ontology and Geographic Kinds, Paper presented to the International Symposium on Spatial Data Handling (SDH'98), Vancouver, Canada, 12-15th July (1998)
17. Smith, B. and D. Mark: Ontology with Human Subjects Testing: An Empirical Investigation of Geographic Categories, American Journal of Economics and Sociology, 58:2, (1999) pp. 245-272

Author Index

Lecture Notes in Computer Science

For information about Vols. 1–3493

please contact your bookseller or Springer